John Gould

The Mammals of Australia

Volume III.

John Gould

The Mammals of Australia
Volume III.

ISBN/EAN: 9783742823557

Manufactured in Europe, USA, Canada, Australia, Japa

Cover: Foto ©berggeist007 / pixelio.de

Manufactured and distributed by brebook publishing software
(www.brebook.com)

John Gould

The Mammals of Australia

THE

MAMMALS OF AUSTRALIA.

BY

JOHN GOULD, F.R.S.,

F.L.S., F.Z.S., M.E.S., F.R.GEOG.S., M.RAY S.; HON. MEMB. OF THE ROYAL ACADEMY OF SCIENCES OF TURIN; OF THE ROYAL
ZOOL. SOC. OF IRELAND; OF THE PENZANCE NAT. HIST. SOC.; OF THE WORCESTER NAT. HIST. SOC.; OF THE
NORTHUMBERLAND, DURHAM, AND NEWCASTLE NAT. HIST. SOC.; OF THE NAT. HIST. SOC.
OF DARMSTADT; OF THE TASMANIAN SOC. OF VAN DIEMEN'S LAND; OF THE NAT.
HIST. SOC. OF STRASBOURG; OF THE NAT. HIST. SOC. OF IPSWICH; AND
CORR. MEMB. SOC. OF NAT. HIST. OF WÜRTEMBERG.

IN THREE VOLUMES.

VOL. III.

LONDON:

PRINTED BY TAYLOR AND FRANCIS, RED LION COURT, FLEET STREET.

PUBLISHED BY THE AUTHOR, 26 CHARLOTTE STREET, BEDFORD SQUARE.

1863.

LIST OF PLATES.

VOL. III.

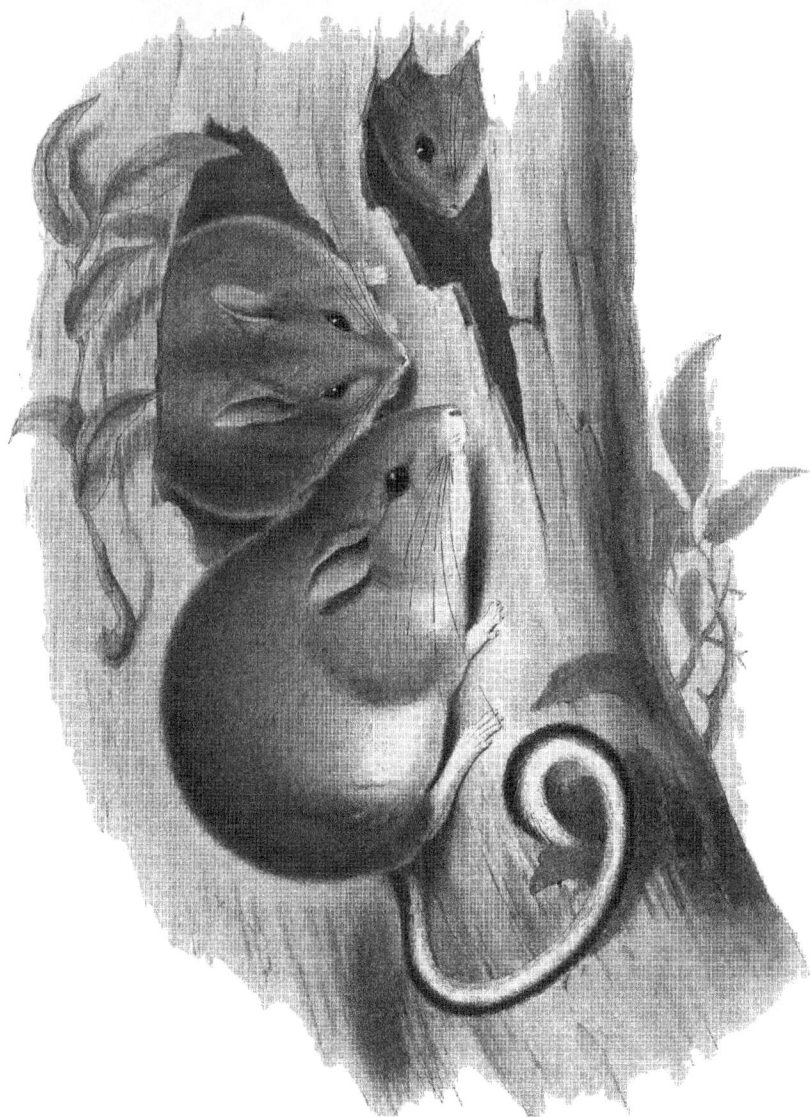

HAPALOTIS ALBIPES, *Licht.*

White-footed Hapalotis.

Hapalotis albipes, Licht. Darst. der Saugth., tab. 29.—Gray, Ann. Nat. Hist., vol. ii. p. 308.—*Id.*, List of Mamm.
in Coll. Brit. Mus., p. 115.—Gould in Proc. Zool. Soc. 1851, p. 126.

Conilurus Constructor, Ogilb. Trans. Linn. Soc., vol. xviii. p. 126.

Bar-roo, Aborigines of the Darling Downs, New South Wales.

The Rabbit Rat of the Colonists, Benn. Cat. of Australian Museum, Sydney, p. 6. no. 30.

THE native habitat of the *Hapalotis albipes* is the south-eastern portions of Australia generally; it is dispersed over all parts of New South Wales, Port Philip and South Australia, but is nowhere very abundant. The South Australian specimens and those of New South Wales assimilate very closely, while those from the Darling Downs district are rather browner in the colouring of the fur and have shorter hind feet. Although I regard this latter animal from the table lands as only a local variety, it may at some future time prove to be distinct.

Judging from my own observations I should say that the *Hapalotis albipes* is strictly nocturnal in its habits, for it sleeps during the day in the hollow limbs of prostrate trees, or such hollow branches of the large *Eucalypti* as are near the ground, in which situations it may be found curled up in a warm nest of dried leaves ; more than once have I, after detecting the animal in its retreat, sawn off the hollow limb and secured it without injury. In a note with specimens from Darling Downs in New South Wales, Mr. Gilbert states that " it is generally found inhabiting hollow logs or holes in standing trees."

The following note respecting this species was sent to me by my kind friend His Excellency Sir George Grey, now Governor of New Zealand, during his Governorship of the Colony of South Australia :—

" This animal lives among the trees. The specimen I send you, a female, had three young ones attached to its teats when it was caught ; the mother has no pouch, but the young attach themselves with the same or even greater tenacity than is observable in the young of the *Marsupiata*. While life remained in the mother they remained attached to her teats by their mouths, and grasped her body with their claws, thereby causing her to present the appearance of a Marsupial minus the pouch. On pulling the young from the teats of the dead mother, they seized hold of my glove with the mouth and held on so strongly that it was difficult to disengage them."

I had frequent opportunities of observing this animal in a state of nature during my rambles in the interior of Australia, and Mr. Gilbert was equally fortunate during his short sojourn in New South Wales. I mention this, because certain habits and nest-making propensities have been referred to this animal by Sir Thomas Mitchell, W. Ogilby, Esq., and others, which belong not to this species, but to the *Hapalotis conditor*, a fact which is fully established by the drawings, specimens and notes of that species made on the spot and communicated to me by Captain Sturt.

Fur long, close and soft ; head, ears, upper surface, flanks and outer surface of the limbs grey at the base and ashy brown on the surface, interspersed with numerous fine black-tipped hairs ; whiskers and a narrow line around the eye black ; under surface of the body, inner surface of the limbs, hands and upper surface of the feet white ; upper surface of the tail dark brown ; sides and under surface white.

The figures are of the size of life.

Hullmandel & Walton, Imp.

HAPALOTIS ALBIPES, *Gould*

J. Gould and H.C. Richter, del. et lith.

HAPALOTIS APICALIS, *Gould.*

White-tipped Hapalotis.

Hapalotis apicalis, Gould, in Proc. of Zool. Soc., 1851, p. 126.

THIS new species is about the size of, and similar in colour to, *H. albipes*, but it differs in having larger ears, much more delicately formed feet, the tail nearly destitute of the long brushy hairs towards the tip, and smaller eyes.

I possess a single example only of this species; it was procured by Mr. Strange in South Australia. There is an animal in spirits in the British Museum, presented by R. C. Gunn, Esq., from Van Diemen's Land, which accords very closely with it in the colouring of the fur, and in the rat-like form of the tail; it is, however, of much smaller size, and in all probability will prove to be a new species.

Face and sides of the neck blue-grey; upper part of the head, space between the ears, the ears and upper parts of the body pale brown, interspersed with numerous fine black hairs; under surface white; flanks mingled grey and buffy white; fore feet white, with an oblique mark of dark brown separating the white from the greyish brown of the upper surface; hinder tarsi and feet white; basal three-fourths of the tail brown, apical fourth thinly clothed with white hairs.

The figures are the size of life.

HAPALOTIS HEMILEUCURA, _Gray_

HAPALOTIS HEMILEUCURA, *Gray*.

Elsey's Hapalotis.

Hapalotis hemileucura, Gray in Proc. of Zool. Soc., part xxv. p. 243.

It is with a degree of mixed pleasure and regret that I bring before the notice of the scientific world this new species of *Hapalotis*. It was brought home by that young and intelligent naturalist, the late Mr. J. R. Elsey, Surgeon to the expedition conducted by A. C. Gregory, Esq., from the north-western coast of Australia to Moreton Bay : all who like myself had an opportunity of becoming acquainted with the amiable qualities of this gentleman, cannot but regret the loss the science of natural history has sustained by his premature decease. On the part of Dr. Gray, I brought this animal before the Meeting of the Zoological Society held on the 24th of November, 1857, and gave it the name of *hemileucura*, a term suggested by the parti-colouring of the tail. Only a single specimen was procured, and this is now in the British Museum. I am unable to state the precise locality in which it was obtained, but believe it was about midway between the Gulf of Carpentaria and Moreton Bay.

The *Hapalotis hemileucurus* is a harsh wiry-furred animal, nearly allied to, but considerably larger than, the *H. melanura*, from which it also differs in having the apical half of the tail white.

Head, all the upper surface and flanks very light sandy brown, with numerous, but thinly placed, fine, long black hairs ; under surface buffy white, with even lighter feet and fore-arms ; tail brown, deepening into black about the middle, beyond which the apical portion is white, the white hairs being prolonged into a small tuft at the tip.

		inches.
Length from the nose to the base of the tail	8
,, of the tail	6¼
,, ,, fore-arm	1¼
,, ,, tarsus and toes	1¼

The figures are of the natural size.

HAPALOTIS HIRSUTUS, *Gould.*

Long-haired Hapalotis.

Mus hirsutus, Gould in Proc. of Zool. Soc., part x. p. 12.—Ib. Ann. & Mag. Nat. Hist., vol. x. p. 405.
Hapalotis hirsutus, Gould in Proc. of Zool. Soc., part xix. p. 127.

THE discovery of this rare Australian animal is due to the researches of the late Mr. Gilbert, who obtained a single specimen during his sojourn at Port Essington on the Cobourg Peninsula in 1840; since that period a second example from the same locality has been sent to this country, and, as well as the former, deposited in the British Museum. It will be seen, by the synonyms above given, that I at first regarded this animal as a true *Mus*, and that I subsequently assigned it a place in the genus *Hapalotis*. I am, however, by no means satisfied that this is its right situation, and think it possible that, when a sufficient number of specimens have been received to justify the formation of a correct opinion upon the subject, it may be found desirable to constitute it the type of a new genus.

The following is a copy of my original description of the animal, published in the 10th Part of the Proceedings of the Zoological Society of London:—

" Fur coarse and shaggy; on the upper parts of the body the shorter hairs are of a yellowish-brown colour, but the longer interspersed hairs, being numerous and of a black colour, give a deep general tint to those parts; the under parts of the body are of a rusty-yellow colour, tinted with brownish on the neck and chest, and having a more decided rust-colour on the abdomen; tail well clothed with lengthened hairs, especially on the apical half, where the scales are hidden by them; those at the point of the tail measure upwards of an inch in length; on this part they have a rusty hue, but on the remaining portions they are black."

The Plate represents the animal of the natural size.

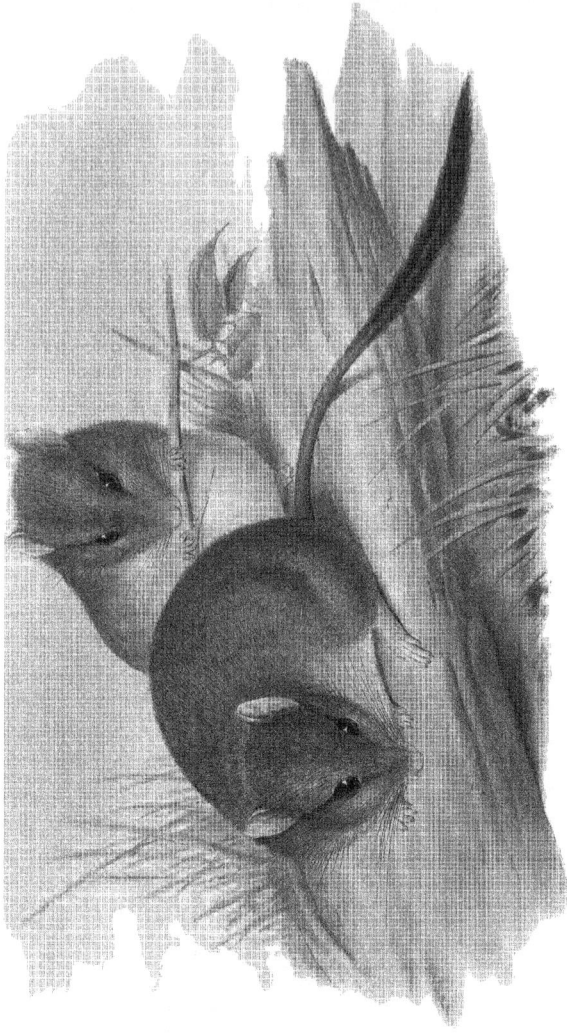

J. Gould and H.C. Richter, del. et lith.

HAPALOTIS PENICILLATA, Gould.

Hullmandel & Walton, Imp.

HAPALOTIS PENICILLATA, *Gould.*

Pencil-tailed Hapalotis.

Mus penicillatus, Gould in Proc. of Zool. Soc., Part X. p. 12.—Ann. and Mag. Nat. Hist., vol. x. p. 405.—List of Mamm. in Brit. Mus., p. 109.

Hapalotis melanura, List of Mamm. in Brit. Mus., p. 115.?

THIS animal was procured by Mr. Gilbert during his short sojourn at Port Essington on the Cobourg Peninsula in Northern Australia, in which part of the country it was also obtained by Mr. MacGillivray and sent by him to the late Earl of Derby; specimens from the same country are also contained in the collection at the British Museum.

It is in every respect a true *Hapalotis*, and may be readily distinguished from the other members of the genus by the blackness of its tail, the hairs of which are much lengthened; and by the rigid, almost spiny, nature of the hairs clothing the back. Its habits would seem to be somewhat singular, inasmuch as it is frequently found among the swamps on the sea-shore; I have no evidence, however, that it is not also found in the interior of the country. I find the following note respecting it among the papers of the late Mr. Gilbert:—

"This little animal is only seen on the beach where there are large *Casuarina* trees, in the dead hollow branches of which it forms a nest of fine dry grass, and retires during the day; in the evening it leaves its retreat and proceeds to the beach, where it may be seen running along at the edge of the surf as it rolls up and recedes, apparently feeding upon any animal matter washed up by the waves."

The fur of the upper surface is greyish brown grizzled with buff, with a rusty tint on the region of the occiput and back of the neck; around the angle of the mouth, the chin, throat, and all the under parts of the body, as well as the feet and inner side of the legs, are white with a faint yellow tint or cream-coloured, and the hair of these parts is of a uniform tint to the roots except on the chest, where they are grey next the skin: the tail is sparingly clothed at the base with minute bristly hairs; but about the middle the hairs become of a black colour and longer, and towards the apex attain a considerable length, measuring at and near the tip half an inch or more: the ears are sparingly clothed with minute hairs.

The figures represent the two sexes of the size of life.

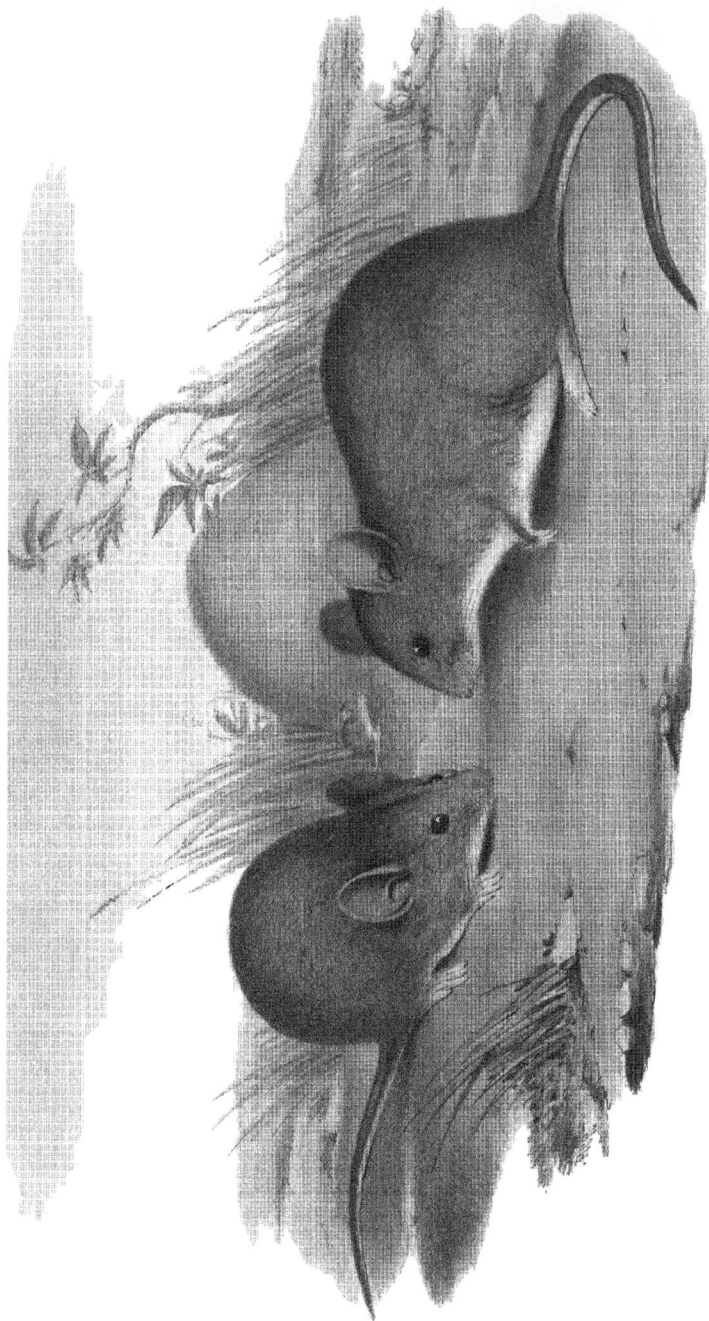

HAPALOTIS CONDITOR, *Gould.*

Building Hapalotis.

Mus conditor, Gould in Sturt's Narr. of Exp. to Central Australia, vol. i. pl. in p. 120 ; vol. ii. App. p. 7.

For a knowledge of this curious little animal, we are indebted to the researches of Captain Sturt, who, during his recent expedition into the central portion of Australia, found it inhabiting the brushes of the Darling ; there is little doubt that it had been previously met with by Major Mitchell, who, in the second volume of his " Three Expeditions into the Interior of Eastern Australia," page 263, when speaking of the specimens collected during the journey, mentions, among others, " the flat-tailed rat from the scrubs of the Darling, where it builds an enormous nest of branches and boughs, so interlaced as to be proof against any attacks of the native dog ;" but as the specimen he procured appears never to have been described, the credit of its first introduction to science is due to the first-mentioned traveller.

In its general form and dentition it is very nearly allied to the members of the genus *Mus,* but its lengthened and broad hind-feet, large ears, and its habit of constructing a nest, are characters which in an equal degree ally it to the *Hapaloti,* with which, upon a closer examination of its structure, I am induced now to associate it.

Captain Sturt states that it " inhabits the brushes of the Darling, but was not found beyond latitude 30°. It builds a nest of small sticks varying in length from three to eight inches, and in thickness from that of a quill to that of the thumb, arranged in a most systematic manner so as to form a compact cone like a bee-hive, about four feet in diameter and three feet high ; those at the foundation are so disposed as to form a compact flooring, and the entire fabric is so firm as almost to defy destruction except by fire. The animal, which is like an ordinary rat, only that it has longer ears and the hind-feet are disproportioned to the fore-feet, lives in communities, and traverses the mound by means of passages leading into the apartments in the centre. One of these nests or mounds had five holes or entrances at the base, nearly equidistant from each other, with passages leading from them to a hole in the ground beneath, in which I am led to conclude they had their store. There were two nests of grass in the centre, with passages running up to them diagonally from the bottom ; the nests were close together, but in separate compartments, with passages communicating from the one to the other."

Fur soft and silky to the touch ; general colour greyish brown, becoming of a darker hue down the centre of the head and back, in consequence of the tips of the hairs being dark brown ; under surface pale buff, the whole of the fur dark slate-grey at the base ; a slight wash of rufous between the ears ; whiskers very long, exceedingly fine, and of a blackish brown hue ; fore-feet brown, hinder feet pale brown ; tail brown above, paler beneath.

The figures are of the natural size.

HAPALOTIS MURINUS, *Gould*

HAPALOTIS MURINUS, *Gould.*

Murine Hapalotis.

Hapalotis murinus, Gould, in Proc. of Zool. Soc., Part xiii. p. 78, and 1851, p. 127.

THE large size of its ears, the peculiar softness of its fur, and the whiteness and length of the hinder feet of this animal induced me some years ago to characterize it under the generic name of *Hapalotis* rather than that of *Mus*, and I still adhere to the opinion I then formed, that it must not be associated with the true Rats. The original specimen from which my description was taken was procured by Mr. Gilbert on the plains bordering the rivers Namoi and Gwydyr, where the natives informed him it was very abundant. Mr. Strange subsequently sent me examples from the neighbourhood of Lake Albert in South Australia, which, although of somewhat larger size, are, I believe, identical. He states that he found them in families on the edge of the small dry salt-water lagoons of the plains, and that they did not appear to go far from their habitations.

Fur remarkably soft and delicate, and of a slate-grey tint next the skin; on the upper surface and the sides of the body the exposed portions of the hairs are of a delicate ochreous yellow, with a considerable admixture of black, the points of the hairs being of that colour; ears tolerably well clothed with small hairs of a white hue, excepting on the fore part of the outer surface, where they assume a dusky greyish tint; under surface buffy white; tail moderately clothed with hairs, but not so thickly as to hide the scales; on the upper surface some of these hairs are white and others blackish, on the sides and under surface of the tail they are pure white; whiskers black at the root, greyish at the point; hands and feet buffy white.

The figures represent the two sexes of the size of life.

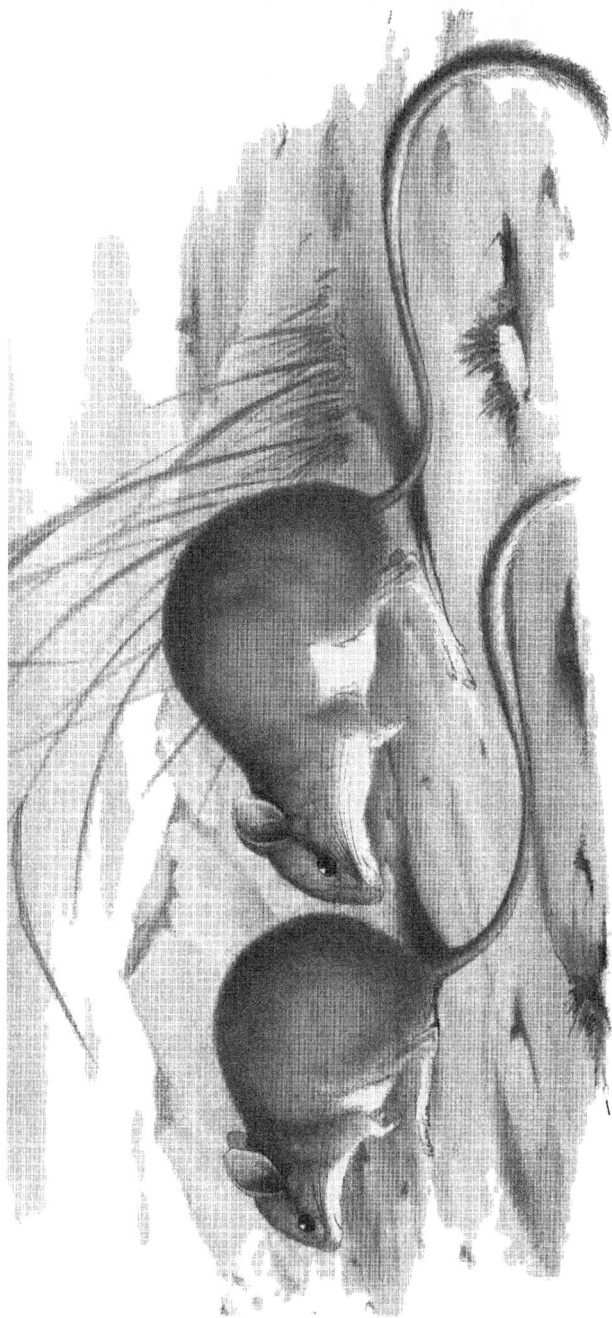

HAPALOTIS LONGICAUDATA. Gould.

HAPALOTIS LONGICAUDATA, *Gould.*

Long-tailed Hapalotis.

Hapalotis longicaudata, Gould in Proc. of Zool. Soc., Part XII. p. 104.

Kor-tung and *Gool-a-wa*, Aborigines in the neighbourhood of Moore's River in Western Australia.

ALTHOUGH very similar in form and style of colouring to the *Hapalotis Mitchellii*, the larger size and the greater length of its tail are characters by which the present animal may be distinguished from that species.

The interior of Western Australia is the only locality in which the *Hapalotis longicaudata* has been procured. The individuals forming the subject of the accompanying Plate were obtained in the vicinity of Moore's River, and now form part of the collection at the British Museum. They were sent to me by Mr. Gilbert, whose notes relative to the present species may not prove uninteresting :—

" This species differs considerably in its habits from the *Djyr-doie-in* (*Hapalotis Mitchellii*), for while that animal burrows in sandy districts, the favourite haunt of the present species is a stiff and clayey soil. It is also very partial to the mounds thrown up by the *Boordee's* (*Bettongia Grayii*) and the *Dal-goitch* (*Peragalea lagotis*). It is less destructive to the sacks and bags of the store-rooms, but, like the *H. Mitchellii,* is extremely fond of raisins."

All the upper surface and the outside of the limbs pale sandy, interspersed on the head and over the back with numerous fine black hairs, which becoming longer on the lower part of the back and rump, give that part a dark or brown hue ; ears naked and of a dark brown ; sides of the muzzle, all the under surface and the inner surface of the limbs white ; tail clothed with short dark brown hairs at the base, with lengthened black hairs tipped with white on the apical half of its length, the extreme tip being white ; tarsi white ; whiskers very long, fine and black ; the fur is close, very soft, and of a dark slaty grey both on the upper and under surface.

The figures represent a male and a female of the natural size.

HAPALOTIS MITCHELLII.

HAPALOTIS MITCHELLII.

Mitchell's Hapalotis.

Dipus Mitchelli, Ogilby in Linn. Trans., vol. xviii. p. 129.—Mitch. Trav., vol. ii. p. 144. pl. 29.—Gould in Proc.
    of Zool. Soc., Part VIII. p. 151.

Hapalotis Mitchellii, Gray, App. to Grey's Trav., vol. ii. p. 404.

————— *Gouldii,* Gray, App. to Grey's Trav., vol. ii. pp. 404 and 413.

Djyr-doub-in, Aborigines around Perth, and

Mät-tee-geteh, Aborigines in the neighbourhood of Moore's River, Western Australia.

———————

THE animal here represented was originally described by Mr. Ogilby under the name of *Dipus Mitchellii,* from a drawing by Major Sir Thomas L. Mitchell of a specimen obtained by him on the banks of the river Murray in South Australia, and now deposited in the Museum at Sydney; since that period specimens have been sent to the Zoological Society of London by the late Mr. J. B. Harvey from South Australia, and to myself by Mr. Gilbert from Western Australia, all of which appear to be identical with the animal discovered by Sir Thomas Mitchell; at least the specimens from Southern and Western Australia have been found on comparison to be precisely similar, and Mr. Gilbert informs me that on examining the Major's specimen in the Sydney Museum, he could perceive no specific difference between it and those transmitted by himself from Western Australia. That they are identical there can be little doubt, when we take into consideration that Sir Thomas Mitchell's specimen was procured at no great distance from the locality in which Mr. Harvey obtained his.

The range of this species is very extensive, and it is probable that the greater portion of the interior of the country will hereafter be found to be inhabited by it.

The only information received respecting the habits of this animal is, that in Western Australia it burrows in the ground; taking up its abode on the sides of grassy hills tolerably well-clothed with small trees growing in a light soil. It occasionally makes its way into the stores of the settlers, and commits depredations on the provisions, particularly sugar and raisins, of which it is exceedingly fond.

The sexes in size and colour offer no material difference.

All the upper surface and the outside of the limbs very pale sandy, interspersed over the head and back with fine black hairs, which becoming numerous and longer on the lower part of the back and rump, give that part a black or brown hue; ears naked and of a dark brown; sides of the face, all the under surface, inner side of the limbs and feet greyish white; down the centre of the throat and chest a broad patch of pure silky white; upper surface of the tail dark brown, under surface white, the hairs becoming much lengthened on the upper surface at the tip; whiskers very long, fine and black; the fur is close, very soft, and of a slaty grey at the base, both on the upper and under surface.

The accompanying Plate represents the animal in three positions, and of the natural size.

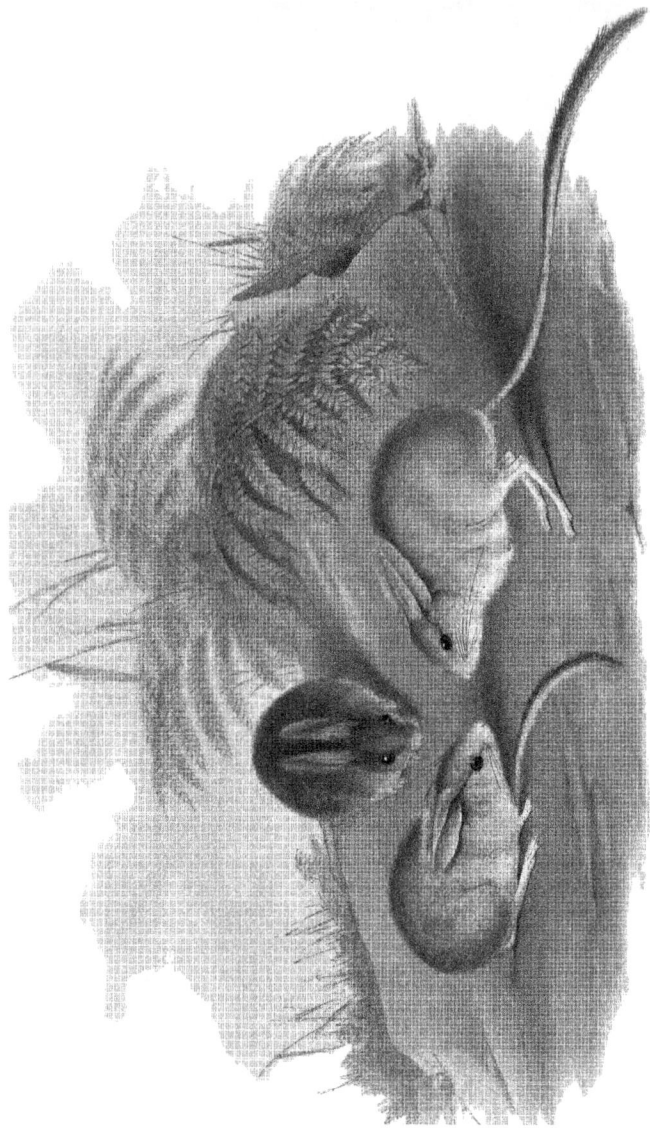

HAPALOTIS ? ERVINUS, Gould.

HAPALOTIS CERVINUS, *Gould.*

Fawn-coloured Hapalotis.

Hapalotis cervinus, Gould, in Proc. of Zool. Soc. 1851, p. 127.

If the Great Red Kangaroo may be extolled as the finest of the Kangaroos, it must be conceded that the present animal is the most graceful and elegant of the Jerboa-like rodents to which the generic term of *Hapalotis* has been applied. For its discovery and introduction to this country we are indebted to the researches of Captain Sturt, who has thus afforded another instance of the anxiety with which this intrepid traveller seeks to promote the cause of science, not only in his immediate vocation as a soldier and explorer, but in the department of zoology, a department never neglected by him whenever he has had opportunities of adding to its stores. It was during the most hazardous of his journeys towards the centre of Australia, that Captain Sturt first met with this pretty species.

"On the 20th," says Captain Sturt, "we found ourselves in lat. 29° 6', and halted on one of those clear patches on which the rain-water lodges, but it had dried up, and there was only a little for our use in a small gutter not far distant. Whilst we were here encamped, a little Jerboa was chased by the dogs into a hole close by the drays, which, with four others, we succeeded in capturing by digging for them. This beautiful little animal burrows in the ground like a mouse, but their habitations have several passages leading straight, like the radii of a circle, to a common centre, to which a shaft is sunk from above, so that there is a complete circulation of air along the whole. We fed our little captives on oats, on which they thrived and became exceedingly tame. They generally huddled together in a corner of their box; but when darting from one side to the other, they hopped on their hind legs, which, like those of the Kangaroo, are much longer than the fore ones, and held the tail perfectly straight and horizontal. At this date they were a novelty to us, but we subsequently saw great numbers of them, and ascertained that the natives frequented the sandy ridges in order to procure them for food. Those we succeeded in capturing were, I am sorry to say, lost from neglect. This species feeds on tender shoots of plants, and must live for many months without water, the situation in which it is found precluding the possibility of its obtaining any for lengthened intervals."

The whole of the head, upper surface and sides of the body of the most delicate fawn-colour, interspersed with numerous fine black hairs on the head and back; whiskers greyish black; nose and under surface white; tail pale brown, lighter beneath; ears very large, somewhat pointed, and nearly destitute of hairs.

The figures are of the natural size; the darker-coloured figure representing a variety sometimes met with.

MUS FUSCIPES, *Waterh.*

Dusky-footed Rat.

Mus fuscipes, Waterh. in Darwin's Zool. of the Voy. of H.M.S. Beagle, Mammalia, p. 66. pl. 25.—Cat. of Mamm.
in Brit. Mus., p. 111.
—— *lutreola*, Gray, App. to Grey's Journ. of Two Exp. of Disc. in Australia, vol. ii. p. 409.

This species of Rat is distributed in abundance over the whole of the southern portion of Australia ;
but I have no evidence that its range extends to the north coast. Specimens from Swan River in Western
Australia, the swamps and thick brushes of New South Wales, the intermediate colony of South Australia,
and the islands in Bass's Straits, differ in no respect from each other. Its favourite haunts are low and
humid situations where long grass and herbage abound, and the banks of freshwater brooks and lagoons

Although belonging to a different genus, it presents in its aquatic habits and in many of its actions a
striking resemblance to the common Water Vole (*Arvicola amphibius*) of Europe ; like that animal, it swims
with the greatest case, and may be constantly seen crossing and recrossing the small brooks and water-
holes so abundant in the localities it frequents. It is rather less than the *Mus Rattus* in size, but is of
a stouter form, and is moreover remarkable for the great length and softness of its fur, and the brown colour
of its feet.

The general tint of the upper surface and the sides of the head and body is blackish brown, with an admixture of grey ; of the under surface greyish white ; the feet are brown, the hairs being greyish at the
tip ; the tail is black, and but sparingly clothed with short bristly hairs ; the cars are rather sparingly clothed
with hairs, which are for the most part of a brownish grey colour ; the ordinary fur of the back is about
three-quarters of an inch in length and very soft, of a deep grey colour broadly annulated with brownish
yellow near, and blackish at, the tip ; the longer black hairs measure upwards of an inch and a quarter in
length ; the incisor teeth are orange-coloured.

The figures in the accompanying Plate represent the animal correctly both in size and colour.

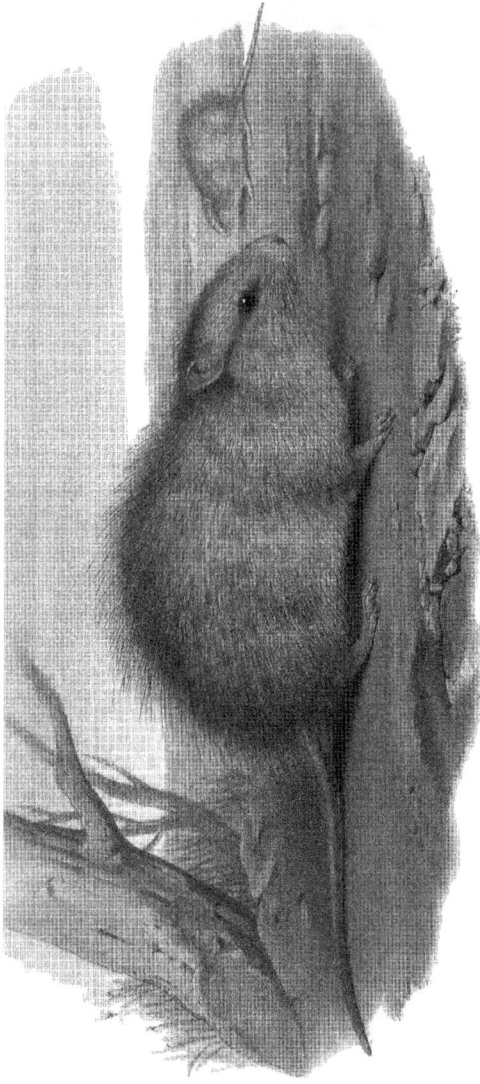

MUS VELLEROSUS, *Gray*

MUS VELLEROSUS, *Gray*.

Tawny Rat.

Mus vellerosus, Gray in Proc. of Zool. Soc., part xv. p. 5.

In the fifteenth part of the "Proceedings of the Zoological Society of London," above referred to, will be found the description of a species of Rat, sent from South Australia by His Excellency Governor Grey. This supposed species received from Dr. Gray the name of *Mus vellerosus* : I say supposed species, because I believe it to be a *lusus*, either of the *Mus fuscipes*, or some nearly allied species ; still, although entertaining this opinion, I have considered it necessary to give an accurate figure of the animal in the present work, and I must leave it to future zoologists to ascertain if it be or be not a true species. It differs from the *Mus fuscipes* not only in its tawny colouring, but in the great length of its furry coat, all the hairs of which are of an equal length, or nearly so ; it is also very different from the *Mus longipilis*, with which indeed I am convinced it has no relationship whatever. Only a single example has yet reached this country, and it is on this that Dr. Gray has founded the species, accompanied with the following remark :—

"This rat has the dentition and somewhat the general appearance of *Mus fuscipes*, Waterh., but the skull and animal are considerably larger, and the fur is very much longer and paler."

Fur long and rather soft to the touch ; general colour reddish brown, varied with whitish interspersed hairs, becoming paler on the sides and still paler beneath, the base of the fur being bluish grey ; feet and tail brown.

The animal is figured on the accompanying Plate of the natural size.

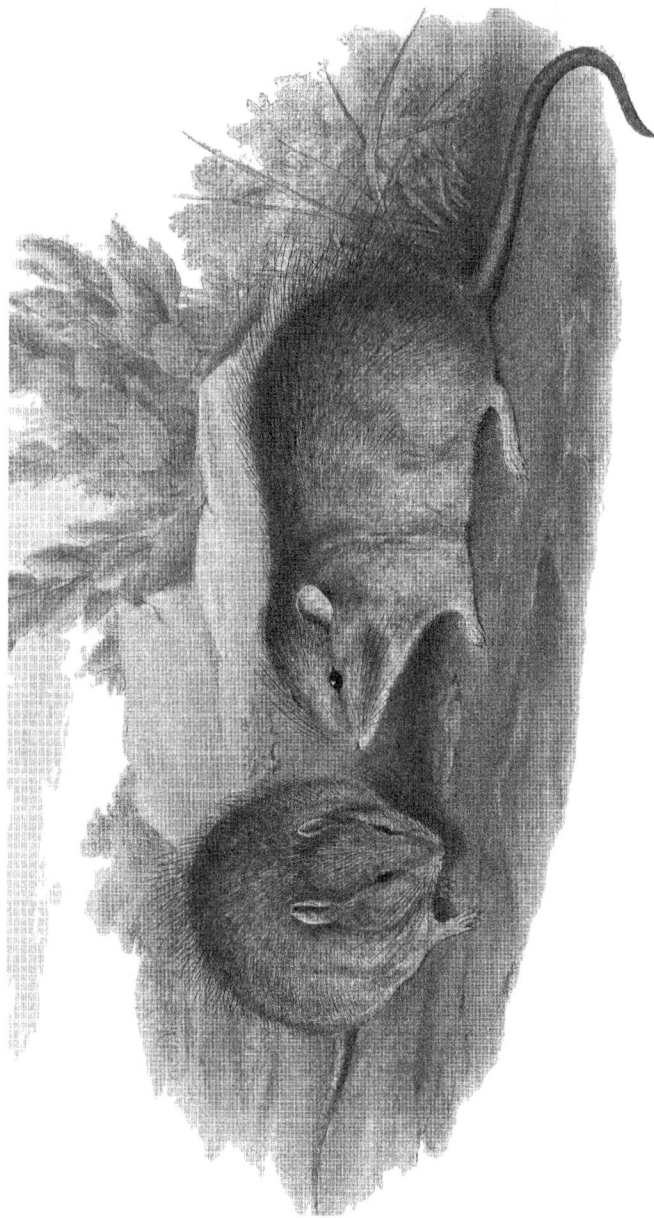

J.Gould and H.C.Richter del et lith.

MUS LONGIPILIS, Gould

Hullmandel & Walton, Imp.

MUS LONGIPILIS, *Gould.*

Long-haired Rat.

I AM indebted to the Directors of the Australian Museum at Sydney for permission to figure this remarkable species of Rat, and for the loan of the unique specimen from which my drawing was taken. In size it approximates very closely to the Common Rat of Europe (*Mus Rattus*), but is at once distinguished from that species by the light buffy hue of its fur, and by the great length of the numerous black hairs interspersed along the back, which latter feature has suggested the specific name of *longipilis*.

In the brief notes kindly transmitted to me by Mr. William Sheridan Wall, that gentleman informs me that it was killed by his late brother, Mr. Thomas Wall, during his expedition to the Victoria River, on a desert which abounded with these animals. "In the absence of vegetation, it was interesting to ascertain, if possible, their means of existence. The stomachs of several were examined with this view, and all were found to contain a fleshy mass, leading to the supposition that they preyed upon each other, for no other animal was found to inhabit the locality." This mode of feeding was doubtless only temporary, probably caused by the entire absence, at the time, of the seeds and other vegetable substances suitable to its economy. It is to be regretted that more examples of this new species were not procured, especially as the one I have figured must be returned to the Australian Museum; examples of so curious a Rat would be very desirable accessions to our national and other collections.

Fur very long, hairy and somewhat harsh to the touch, of a greyish brown at the base, and tawny buff on the surface, numerously interspersed, especially along the back, with very long, fine, black hairs; under surface of the body buffy grey; feet flesh-colour, sparingly clothed with silvery white hairs; tail thinly beset with fine, stiff, black hairs, between which the usual scaly appearance is perceptible.

Total length, from the tip of the nose to the end of the tail, $13\frac{1}{4}$ inches; of the tail, $5\frac{3}{4}$; of the nose to the ear, $1\frac{1}{4}$; of the ear, $\frac{3}{4}$; of the tarsi, $1\frac{1}{16}$ inch.

The figures are of the natural size.

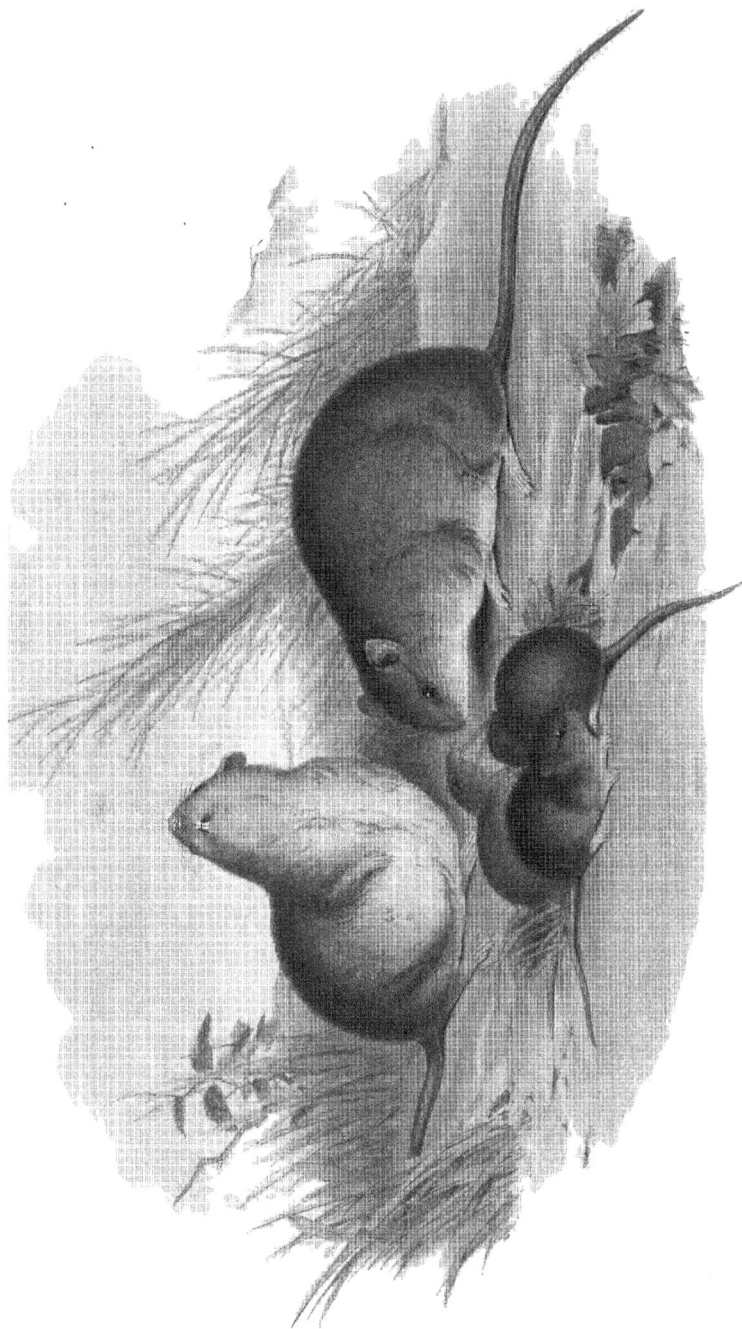

MUS CERVINIPES, *Gould.*

MUS CERVINIPES, *Gould.*

Buff-footed Rat.

Mus cervinipes, Gould in Proc. of Zool. Soc., 1852.

THE species of Rat figured on the accompanying Plate, which is rather widely dispersed over the eastern coast of New South Wales, possesses characters which distinguish it from all the known members of the genus inhabiting that country; its short, soft, adpressed, furry coat, destitute of any lengthened hairs along the back and sides of the body, is one of the characters alluded to, the nearly uniform rufous colouring of its upper surface is another, and its slender, hairless, reticulated tail forms a third. The eastern brushes generally from the River Hunter to Moreton Bay are known to be inhabited by it; but how far its range may extend to the northward is as yet unascertained. Among the numerous specimens sent to me by Mr. Strange, several are labelled with the localities in which they were killed,—viz. Stradbrook Island, Moreton Bay, where it is called *Corrill* by the natives,—Richmond River, where the Aborigines term it *Cunduoo,* —and the plains bordering the upper parts of the River Brisbane.

The specific name has been suggested by the fawn-like colouring of its broad tarsi and feet.

Head, all the upper surface and flanks sandy brown, the base of the fur being dark slate-grey; tarsi and feet fawn-colour; under surface mottled buffy white and grey, the base of the fur being grey, and the extremity buffy white; tail purplish flesh-colour.

In some specimens the buffy white hue predominates and becomes conspicuous on the throat and breast.

In the young animal the upper surface is bluish grey and the under surface greyish white.

The figures in the accompanying Plate represent an adult of each sex and three very young individuals, all of the natural size.

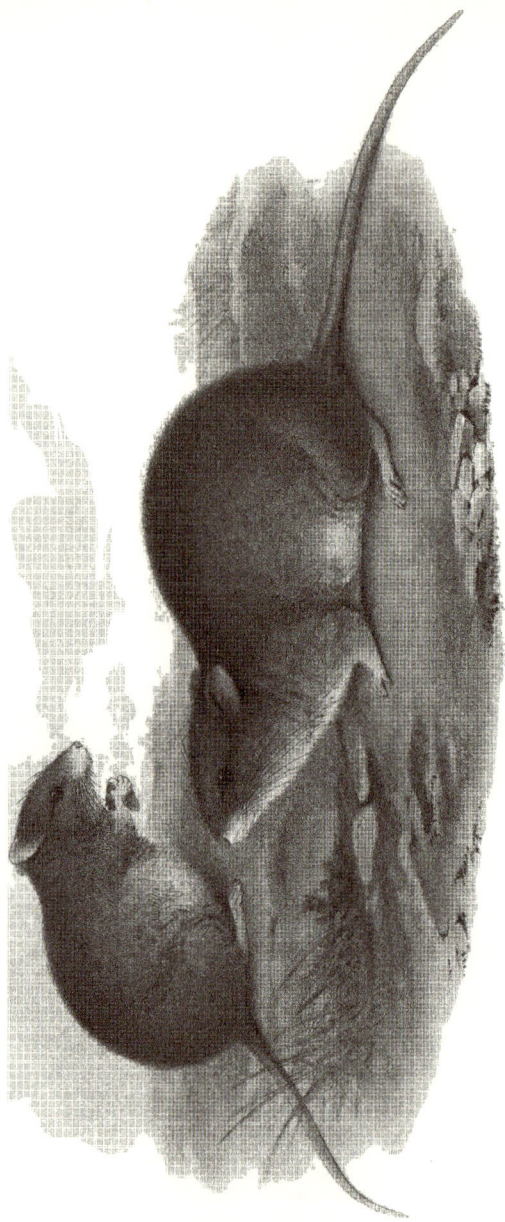

MUS ASSIMILIS, Gould

MUS ASSIMILIS, *Gould.*

Allied Rat.

Mus assimilis, Gould in Proc. of Zool. Soc., part xxv. p. 241.
Moor-deet, Aborigines of King George's Sound.

THE Allied Rat is somewhat numerous in New South Wales. The two specimens from which the characters of the species were taken for the "Proceedings of the Zoological Society," above quoted, were procured by the late Mr. Strange on the banks of the Clarence. I have three other specimens collected by Mr. Gilbert at King George's Sound, which differ only in being about a fifth smaller in all their admeasurements : it is just possible that it will hereafter be found that these latter animals are distinct from the former, but at present I regard them as identical, and if such be the case, the range of the species extends along the whole southern seaboard of the continent from east to west.

The *Mus assimilis* is about the same size as the *Mus decumanus* of Europe, and has a very similar aspect ; its hair, however, is more soft and silky, and its incisor teeth very long and narrow.

Face, all the upper surface and sides light brown, very finely pencilled with black ; under surface greyish buff ; the base of the fur all over the body dark slaty grey ; whiskers black ; tail nearly destitute of hairs ; all the feet clothed with very fine silvery-white hairs, giving those organs a very delicate appearance.

	inches.
Length from the nose to the base of the tail	7¼
,, of the tail	6
,, ,, fore-arm . . .	1
,, ,, tarsus and toes	1¼

The figures are of the natural size.

MUS MANICATUS, *Gould*

MUS MANICATUS, *Gould.*

White-footed Rat.

Mus manicatus, Gould in Proc. of Zool. Soc., Part xxv. p. 242.

THE *Mus manicatus* is a remarkable species of Rat, of nearly the same colour and size, and of a similarly delicate structure, as the well-known Black Rat of the British Islands (*Mus Rattus*), but from which it differs in having the tip of the nose, the front part of the lips, a longitudinal stripe on the breast, and the fore- and hind-feet white, which latter peculiarity suggested the specific appellation of *manicatus* or " gloved."

The only specimen I have yet seen of this animal was procured at Port Essington, on the north coast of Australia, and was subsequently presented to me by J. B. Turner, Esq.

Head, ears, and all the upper surface black, gradually passing into the deep grey of the under surface; nose, fore part of the lips, stripe down the centre of the throat and chest, fore- and hind-feet white; whiskers deep black; tail denuded of hairs.

		inches.
Length from nose to base of tail	. . .	7
,, of tail . .	. . .	5
,, ,, fore-arm . .	. .	1½
,, ,, tarsi and toes .	. .	1¼

The figures are of the natural size.

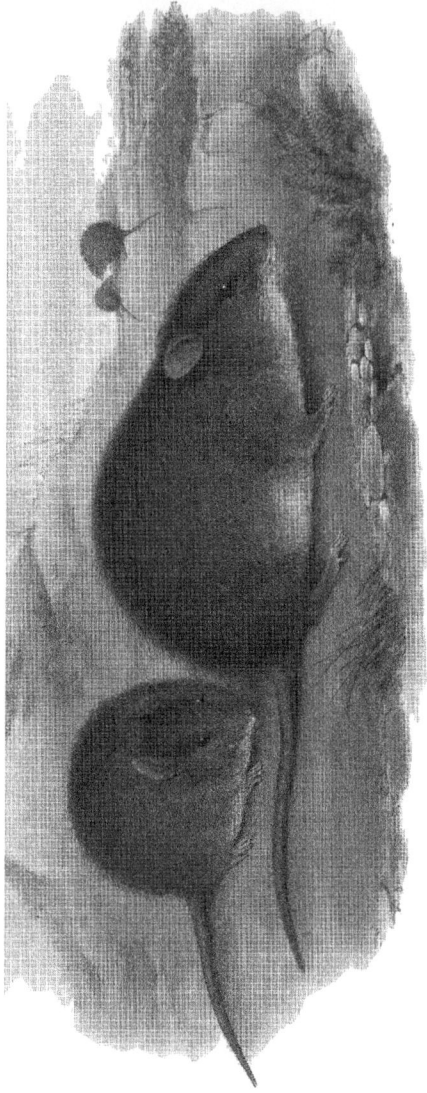

MUS SORDIDUS, *Gould*

MUS SORDIDUS, *Gould.*

Sordid Rat.

Mus sordidus, Gould in Proc. of Zool. Soc., part xxv. p. 242.
Dil-pea of the Aborigines of New South Wales.

VERY fine examples of this robust and compact Rat were procured by the late Mr. Gilbert on the Darling Downs in New South Wales. At present these specimens are in my own collection, but when this work is completed, they will form part of the rich stores of natural history at the British Museum. Mr. Gilbert states that it is common on the plains, and is occasionally found on the banks of creeks, and adds, that it mostly feeds on the roots of stunted shrubs.

The *Mus sordidus* is nearly equal in size to the common Water Vole of England (*Arvicola amphibius*), but it is rather smaller than the *Mus fuscipes* of Australia. It is in every respect a true *Mus* : its incisor teeth, when compared with those of *M. assimilis*, are broad and less elongated, its hair also is coarser and more wiry. Its colouring is as follows :—

Head, all the upper surface, and flanks clothed with a mixture of black and brown hairs, the former hue prevailing along the centre of the back, and both nearly equal in amount on the flanks ; whiskers black ; under surface greyish buff ; hind-feet silvery grey ; fore-feet greyish brown ; tail thinly clothed with extremely fine black hairs.

	inches.
Length from the nose to the base of the tail	6¼
,, of the tail	5
,, ,, the fore-arm	¾
,, ,, the hind-leg and toes . . .	1½

The name of *sordidus* has been assigned to this animal from the dark colouring of its upper surface. The figures are of the natural size.

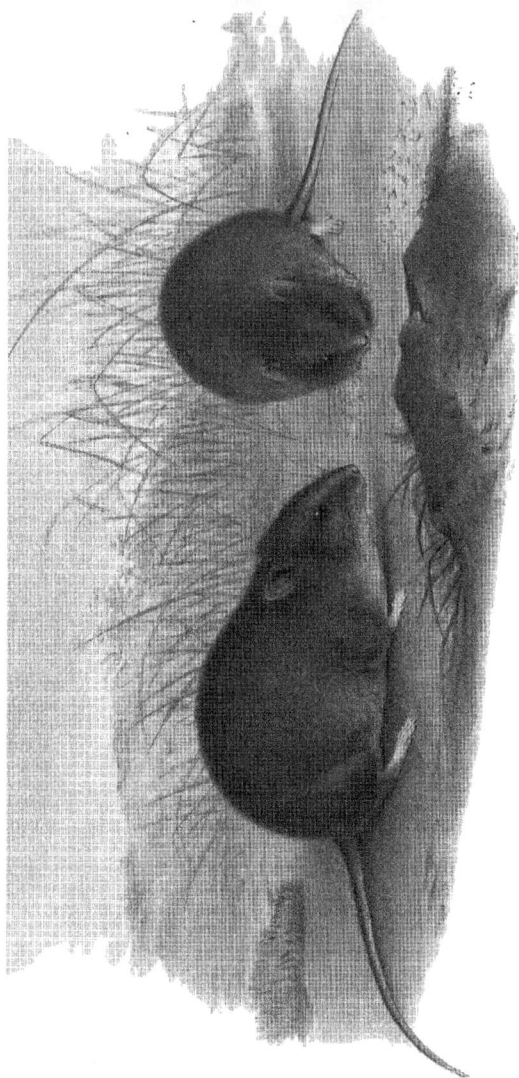

MUS LINEOLATUS, *Gould.*

Plain Rat.

Mus lineolatus, Gould in Proc. of Zool. Soc., part xiii. p. 77.

THIS species of *Mus* was discovered by Mr. Gilbert on the Darling Downs, where it appears to be abundant. In size it is just intermediate between a Rat and a Mouse, taking our own well-known animals for comparison. Two fine examples are now in my own collection, but will hereafter be added to the stores of the National Museum, where they will be at all times accessible to the mammalogist who may wish to investigate this intricate group of animals ;—I say intricate, because so great a sameness of colouring prevails among the species, that it is exceedingly difficult to distinguish one from another; indeed it can scarcely be effected without reference to the specimens themselves ; for, although the utmost care is always taken to secure the accuracy of my illustrations, the minute characters which distinguish them cannot be rendered sufficiently apparent in a drawing.

The dark colouring of the upper surface of the well-clothed tail, contrasted with the light hue of its under portion, are the points which distinguish this species.

Mr. Gilbert states that it is called *Yar-lie* by the natives of the Darling Downs ; that it is common in all the open parts of the grassy plains, and that he believes it is confined to the interior of the country.

The fur of this animal is long and very soft ; on the back the hairs are of a deep slate-grey, but with the exposed portion of a dirty yellowish hue, the points however being black ; long interspersed black pointed hairs are abundant on the back, and give a deep general tint to that part ; on the sides of the body the prevailing tint is greyish-yellow, and the under parts are grey-white, faintly suffused with yellowish ; the hairs on these parts are however of a deepish grey, excepting at the points ; the hairs of the moustaches are rather small and black ; the eye is encircled with black ; the ears are of moderate size and well covered with minute hairs ; those on the outer side are black, excepting on the hinder part, where they assume a greyish-white tint, like those on the inner side of the ear. The feet are rather small and white ; the fore-feet are however greyish at the wrist, and the tarsi are indistinctly suffused with yellowish. The tail is about equal in length to the head and body taken together, well clothed with smallish hairs, which do not however perfectly hide the scales ; those on the upper surface are chiefly brownish-black, but slightly pencilled with whitish in parts ; on the sides and under part they are white.

The Plate represents the animal of the natural size.

MUS GOULDI, *Waterh.*

MUS GOULDI, *Waterh.*

White-footed Mouse.

Mus Gouldii, Waterh. Zool. of Voy. of Beagle, Mamm., p.   . pl. 32. fig. 18, teeth.—Gray, List of Mamm. in Coll. Brit. Mus., p. 111.

—— *Greyii*, Gray in Grey's Journ. of Discoveries in Australia, App. vol. ii. p. 410.

Kurn-dyne, Aborigines of the neighbourhood of Moore's River, in the interior of Western Australia.

THE *Mus Gouldi* is a very distinct and well-marked species, of a size intermediate between that of a Rat and a common Mouse, and may be at all times distinguished by its lengthened, slender, and, white hind feet. It evinces a preference for the plains and sand-hills of the interior, and, as I have seen specimens from the Liverpool Plains, from South Australia, and from the neighbourhood of Moore's River, in Western Australia, appears to range across the southern part of the continent from east to west. The original example from which Mr. Waterhouse took his description was probably from Mr. Coxen's collection, made either on the Upper Hunter or on the interior side of the Liverpool range. Two others transmitted by Mr. Strange were said to have been found between the River Courong and Lake Albert, "and to make their burrows under bushes." Mr. Gilbert states that in Western Australia the animal inhabits the sides of grassy hills where the soil is loose; that its burrows, which are constructed about six inches below the surface, are often of great extent, and that it is generally found in small families of from four to eight in number, inhabiting the same burrow, and even the same nest of dried soft grasses.

Fur soft; general hue buffy-brown, interspersed on the head, upper surface and sides, but particularly on the back, with numerous somewhat longer black hairs; under surface pale buffy-white, washed with a deeper tint of buff on the cheeks and lower portion of the sides; whiskers black; hands and feet white; tail brown above, paler beneath.

The figures are of the natural size.

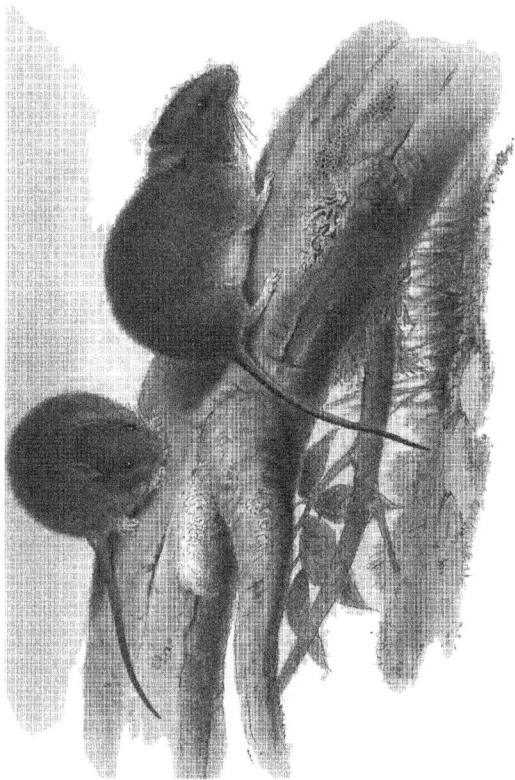

MUS NANUS, *Gould*

MUS NANUS, *Gould.*

Little Rat.

Mus nanus, Gould in Proc. of Zool. Soc., part xxv. p. 242.
Jib-beetch, Aborigines of Moore's River in Western Australia.

THE *Mus nanus* is a very diminutive Rat, with coarse hair and a somewhat short tail; it is even smaller in size than the *Mus Gouldi* and *M. gracilicaudus*, but is more nearly allied to the latter than to any other. Three or four specimens, all of the same size, are contained in the collection at the British Museum, and there are others in the Derby Museum at Liverpool; some of these were collected by Mr. Gilbert on the banks of Moore's River, and the others on the Victoria plains in Western Australia.

Head, all the upper surface, flanks, outer sides of the limbs, and hairs clothing the tail brown, with numerous interspersed fine black hairs; under surface greyish white, becoming much lighter, and forming a conspicuous patch immediately beneath the tail; whiskers black; feet light brown; base of the whole of the fur bluish grey.

		inches.
Length from the nose to the base of the tail	. . . . . .	4
,, of the tail . .	. . . . .	3¼
,, ,, fore-arm . .	. . . .	¼
,, ,, tarsus and toes .	. . .	¾

The figures are of the natural size.

MUS ALBOCINEREUS; Gould.

J. Gould and H.C. Richter, del. et lith.

Hullmandel & Walton Imp.

MUS ALBOCINEREUS, *Gould.*

Greyish-white Mouse.

Mus albocinereus, Gould in Proc. of Zool. Soc., Part XIII. p. 78.

Noô-jee, Aborigines of Perth, Western Australia.

Jûp-pert, Aborigines of Moore's River in the interior of Western Australia.

As yet we have only seen this pretty little Mouse from Western Australia, where it inhabits the sandy districts bordering the sea-shore, particularly those at the back of the sand-hills near the beach a few miles to the northward of Fremantle; in such situations it forms burrows nearly three feet beneath the surface, with two or more openings, one of which is apparently used for no other purpose than that of bringing out the sand, when it becomes necessary to extend the burrow, and this hole is in general nearly filled up by the sand rolling down from the heap. I regret to say that the above meagre account is all that is yet known respecting it; at present its range appears to be very restricted, but future research will doubtless prove that it extends over all parts of Western Australia, wherever suitable localities occur. The remarkable similarity of the colouring of many animals to that of the soil they inhabit, has often been noticed, and the present is another instance of this curious law, which doubtless tends much to enable these little defenceless animals to elude the attacks of their natural enemies; for no two objects so dissimilar in character can be more alike in hue than are the fur of the *Mus albocinereus* and the sandy districts of Western Australia.

This Mouse is rather larger than the Common Mouse of Europe (*Mus musculus*), and its body is considerably stouter in proportion; the head is large; the ears moderate, or perhaps they may be described as rather small; the tail is nearly equal to the head and body in length; the tarsi are very slender: the fur is very long and very soft, and its general hue is pale ashy grey; on the hinder part of the back is a slight brownish tint, produced by a very fine and indistinct pencilling of dusky or pale greyish yellow; the lower part of the sides of the body and the whole of the under parts are white, but not quite pure, having a faint greyish hue; the head is grey-white, pencilled with black; the sides of the muzzle white; the ears are well-clothed with minute greyish white hairs; the feet are white, and if we except some scattered blackish hairs on the upper surface, the tail is also white.

The figures are of the natural size.

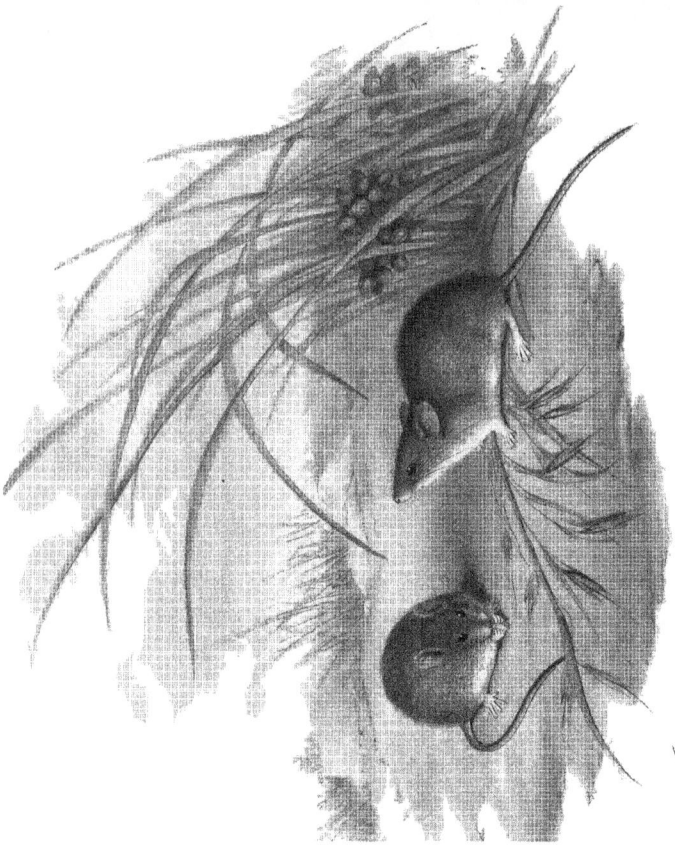

MICE NON-DELAYER. (Plate)

MUS NOVÆ-HOLLANDIÆ, *Waterh.*

New Holland Field Mouse.

Mus Novæ-Hollandiæ, Waterh. in Proc. of Zool. Soc., part x. p. 146.—Gray, List of Mamm. in Coll. Brit. Mus., p. 112.

It is very generally believed that all, or nearly all, the Mammals of Australia are marsupial; but this is not the case; one order at least—the Rodentia—being as fairly represented in that country as in any other. Both rats and mice are in abundance, but they are specifically distinct from those of the northern hemisphere.

The *Mus Novæ-Hollandiæ* inhabits the plains and stony ridges of New South Wales, both in the districts between the mountain-ranges and the sea and in those of the interior. Mr. Waterhouse took his description of this species from an example collected at Yarrundi, on the Upper Hunter, and I have now before me additional specimens from the same district, and others collected on the banks of the river Gwydir, where they were procured by Mr. Gilbert. The animal described by Mr. Waterhouse was, I believe, somewhat immature; his measurements, therefore, will not answer for the adult, which is represented on the accompanying Plate of the natural size. I usually found this species among stones, or under flat slabs of bark, left by the aborigines at their encampments; but Mr. Gilbert states that, while travelling among the high grass in the neighbourhood of the Gwydir, he constantly started it from out of the fissures in the dry ground.

Mr. Waterhouse states, that it approaches most nearly to the *Mus sylvaticus* in form and colouring, but that the tail is considerably shorter than in that animal; he remarks that it also approaches that species in the form of the skull, but has the nasal portion shorter; the molar teeth are of the same structure, but apparently rather larger in proportion.

The fur is rather long and very soft; on the upper parts the hairs are of a deep grey, tipped with brownish-yellow; on the belly the hairs are of a paler grey next the skin, and white externally; the tarsi are rather long and slender; the tail is white beneath and dusky above.

The figures are of the size of life.

MUS DELICATULUS, *Gould.*

Delicate-coloured Mouse.

Mus delicatulus, Gould in Proc. of Zool. Soc., part x. p. 13.—Ann. and Mag. Nat. Hist., vol. x. p. 406.—Gray,
List of Mamm. in Coll. Brit. Mus., p. 112.

Mo-lyne-be, Aborigines of Port Essington.

THE contour and general colouring of this, the smallest and most beautiful species of *Mus* yet discovered in the great country of Australia, strongly remind one of the pretty little harvest mouse, *Mus messorius,* of our own islands. It is a native of Port Essington, where it was discovered by the late Mr. Gilbert, and all we know respecting it is comprised in the following brief notice of it in his Journal :—

"I only met with this species on one occasion, on the Native Companion plains near Point Smith, at the entrance of the harbour, when I found four in a hole which ran along a few inches below the surface for about five feet in a zigzag manner, and terminated in a circular space, wherein was a nest of fine dried grass, in which I captured them."

Two specimens of this little animal are in the collection at the British Museum. Mr. Gray states that I had attached the MS. name of *albirostris* to them; but that appellation not having been published, the term *delicatulus,* under which the animal was certainly described in the "Proceedings of the Zoological Society," is the one retained.

The fur is soft and short; that on the upper parts of the body is of a pale yellow-brown ; the sides are of a delicate yellow tint; and the lower part of the sides of the muzzle, chin, throat, under surface and feet are pure white; on the throat and along the mesial line of the abdomen, the hairs are of a uniform colour to the base ; ears small; feet delicate ; tail slender, and nearly as long as the head and body.

The figures are of the natural size.

HYDROMYS CHRYSOGASTER, *Geoff.*

Golden-bellied Beaver-Rat.

Hydromys chrysogaster, Geoff. Ann. Mus., tom. vi. p. 81. tab. 36. fig. A.—Gray, List of Mamm. in Coll. Brit. Mus.,
p. 121.

The first specimens of Mammalia transmitted to Europe from Australia after the discovery of that country were from New South Wales and Van Diemen's Land; and among others the present animal attracted, at a very early period, the attention of the French naturalists, one of whom—the celebrated Geoffroy St. Hilaire—assigned to it the name of *Hydromys chrysogaster,* a name very suitable to the animal from the localities above mentioned; similar Beaver-Rats are, however, universally spread over the whole of the southern portion of Australia, including the eastern and western confines of that continent; but the animals from each of these localities appear to me to be distinct species; and a most complete series of the whole of them being now before me, I think I shall be able to point out, in my account of each of them, characters of sufficient importance to be regarded as specific.

The present species may be distinguished from all its congeners by the bright golden colouring of the sides of the face, lips, throat, shoulders, flanks and belly, the darker colouring being confined to the crown of the head and the upper part of the back only, whereas in two of the other species this dark colouring occupies so much of the upper part of the body as to include the shoulders and part of the fore arm; and, however near the whole of the species may assimilate in size, the present is the largest, as well as the one in which the colouring is the most contrasted and brilliant.

The native habitat of the *Hydromys chrysogaster* is New South Wales and Van Diemen's Land. It is strictly fluviatile in its habits, frequenting the muddy sides of creeks and water-holes, and the banks of the larger rivers and inlets of the sea. Rather shy in its disposition and nocturnal in its movements, it is not so often seen as might be supposed; at the same time it is by no means difficult to be procured when such an object is desirable. As might be inferred from the structure of its hind feet, the water is its native element; it swims and dives with the greatest facility, and easily secludes itself from view amidst the sedges lining the water's edge, or by descending to its hole after the manner of the common Water Vole of Europe. Like many other of the Australian Mammals, it reposes much on its hinder legs, in which position it may frequently be seen on large stones, snags of wood, or any other prominence near the water's edge.

Head, ears, back, outer surface of the hinder limbs, the portion of the body posterior to them and the base of the tail mingled black and buff, the former hue predominating; sides of the face, of the body, all the under surface and the inner side of all the limbs rich deep reddish orange; outer surface of the arms deep brown; upper surface of hinder feet pale glaucous buff, passing into brown on the tips of the toes; basal half of the tail black, apical half white.

The figures are somewhat smaller than life.

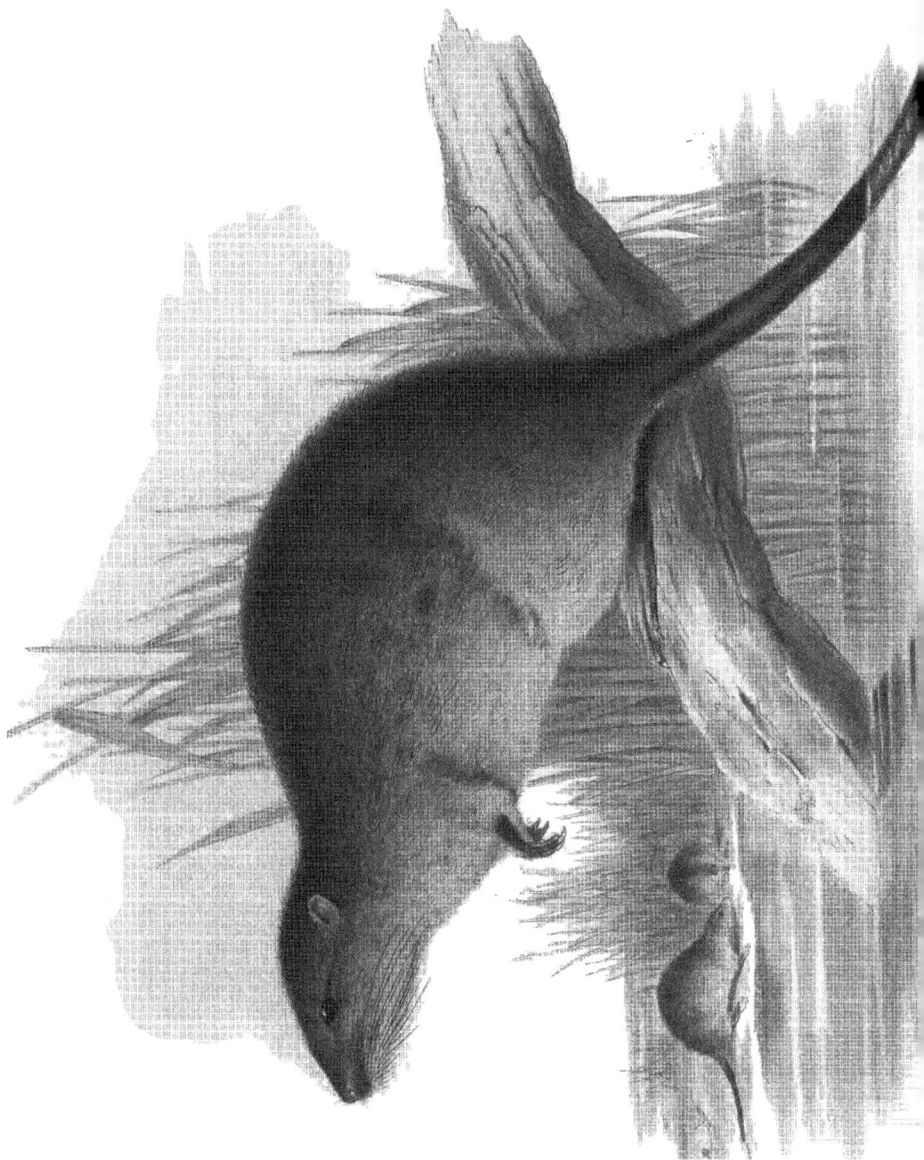

HYDROMYS FULVOLAVATUS, Gould.

Fulvous Beaver-Rat.

I possess specimens of this animal from three different localities—some obtained near the River Murray by E. J. Eyre, Esq., now Lieut.-Governor of New Zealand, others procured at Lake Albert by Mr. Strange, and others shot by myself in the pools of the upper part of the River Torrens, all of which closely resemble each other, but differ very considerably from either of the foregoing species; I have therefore been induced to regard them as specifically distinct. To the Western Australian animal (*H. fuliginosus*) they are allied in the extreme tip of the tail only being white, and to *H. chrysogaster* in the colouring of the under surface, but in no other respect so far as colour goes. As the specific name implies, the whole of the body is washed with golden orange, a tint only relieved by the interspersion of numerous black hairs over the upper surface, giving that part a darker hue, without any decided line of demarcation separating the colouring of the upper from that of the under surface. The habits and economy of this species offer a close resemblance to those of *H. chrysogaster*. I usually found it on the muddy banks of the water-holes of South Australia, where, like the European Water Vole, it lived upon vegetables, mollusks, and other lacustrine animals common to such situations.

The feet of this species are somewhat darker coloured than those of *H. chrysogaster*. The general hue of the fur orange buff, but the numerous black hairs which are dispersed over the head and upper surface give those parts a dusky hue; the whiskers, which in the other species are entirely black, are here mingled black and white; outer surface of the limbs dark brown; upper surface of hinder feet pale brown, deepening into a darker hue on the toes; nails white; tail black, except at the extreme tip, which is white.

The figures are rather under the natural size.

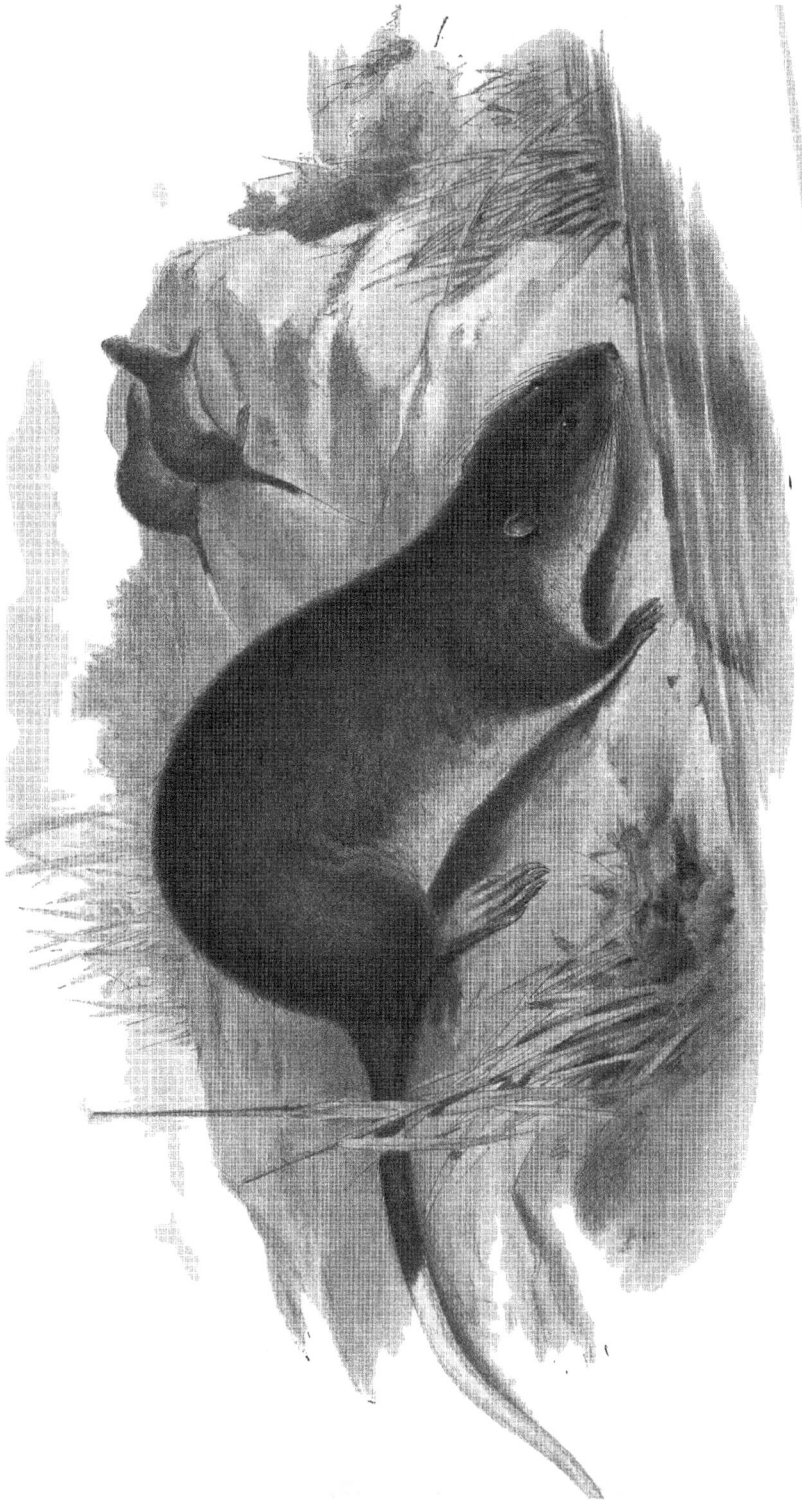

HYDROMYS LEUCOGASTER. Geoff.

HYDROMYS LEUCOGASTER, *Geoff.*

White-bellied Beaver-Rat.

Hydromys leucogaster, Geoff. Ann. Mus., tom. vi. p. 81. tab. 36. figs. B, C, D?

M. GEOFFROY ST. HILAIRE has given the name of *leucogaster* to an animal of this genus, and I believe the subject of the accompanying Plate to be the one to which it was applied. The several specimens contained in my collection were obtained on the banks of the Hunter, Clarence, and other rivers traversing the districts lying between the mountain ranges and the sea. They are all similarly marked, and, as will be seen on reference to the Plate, differ very considerably from *H. chrysogaster*, the tawny white of the under surface occupying the belly only, while the shoulders and upper part of the fore arm are included in the darker colouring of the upper surface. I mention this latter point more particularly, because, were the colouring of the under surface the only difference between the two animals, some persons might suppose that difference to be due to the action of light, which, having abstracted the rich orange colouring, had left the parts thus coloured of a dull or tawny white.

The hinder feet of the two animals also differ, those of *H. leucogaster* being smaller and of a darker colour than those of *H. chrysogaster*, and having the toes of the fore feet for half their length from the nails white, a feature I never observed in the latter species. The only character in which they are alike consists in the extent of the white on the tail, which occupies the terminal half in both species.

Head, all the upper surface, shoulders, sides, outer surface of all the limbs, and the portion of the body posterior to them, mingled black and buffy grey, the former hue predominating; face, all the under surface of the body and the inner side of the limbs buffy white; upper surface of the hinder feet deep purplish buffy white; basal half of the tail black, apical half white.

The figures are somewhat less than the size of life.

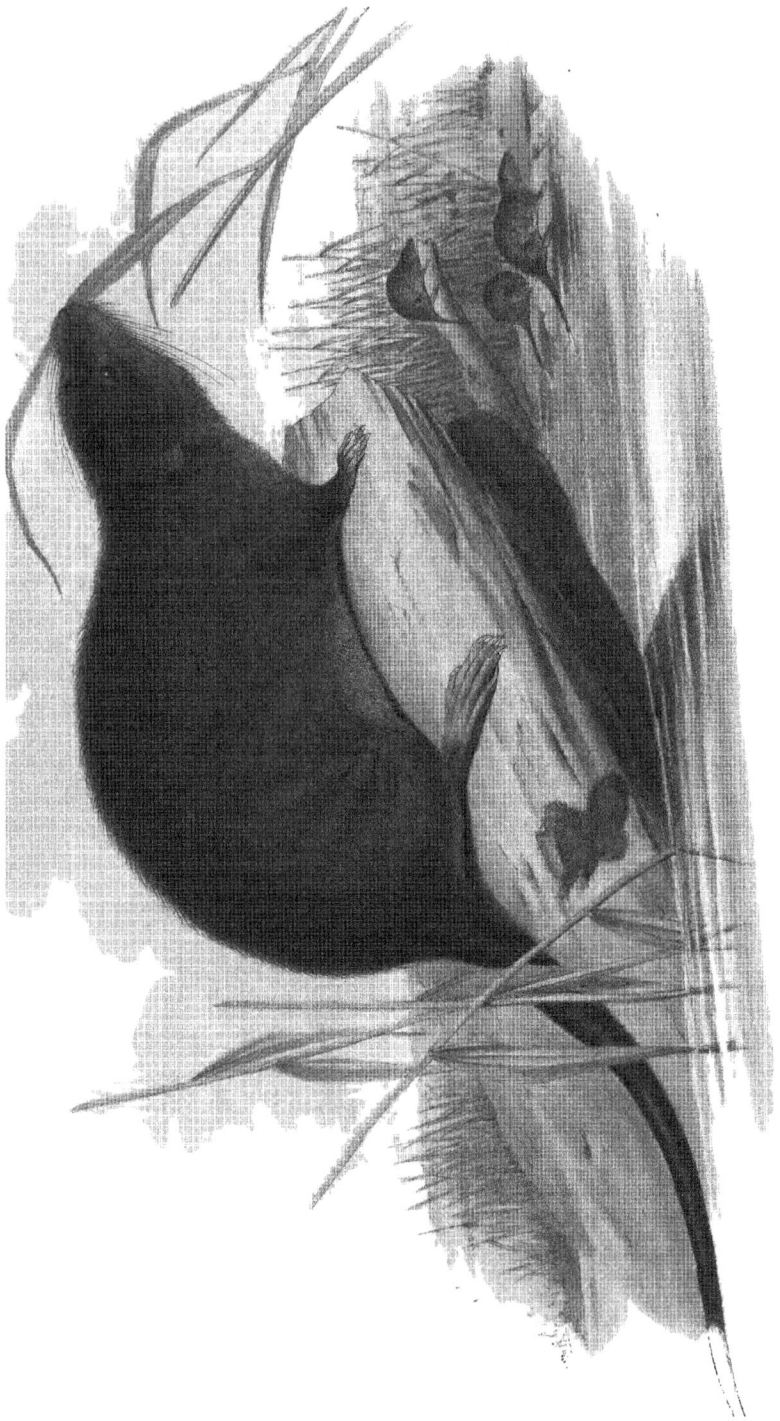

HYDROMYS FULVOGASTER

HYDROMYS FULIGINOSUS, *Gould.*

Sooty Beaver-Rat.

Ngoôr-jou, Aborigines of Perth, Western Australia.
Ngwîr-ri-gin, Aborigines of King George's Sound.

THE specimens of this animal in my collection were procured in the neighbourhood of the lakes near Perth and at King George's Sound in Western Australia, by the late Mr. Gilbert, who in his letter to me on the subject expressed his opinion that they were quite different from any other species he had seen ; and surely an animal so different from all its congeners in the colouring of the body, in the darker colour of the hinder, and the greyish white hue of the fore feet, may with propriety be considered as specifically distinct. Independently of these differences, I may mention, that the uniformity of the body tint is all but unbroken, the face, the centre of the back, and the basal portion of the tail being simply a trifle darker than the rest of the fur. Mr. Gilbert hints that the *H. chrysogaster* also inhabits Western Australia, but in this I believe he was mistaken ; in all probability the South Australian species, to which I have given the name of *fulvolavatus,* is the animal he saw, but failed to procure.

Fur of the upper surface mingled buffy brown and black, the latter hue predominating and producing a deep sooty appearance, especially along the back, whence the specific name ; whiskers and fur of the face black ; that of the tail is also black, except the apical inch and a half, which is white ; fur of the under surface pale greyish brown ; fur of the outer surface of the limbs dark brown, of the fingers white ; nails white.

The figures are rather less than the size of life.

PTEROPUS POLIOCEPHALUS, *Temm.*

PTEROPUS POLIOCEPHALUS, Temm.

Grey-headed Vampire.

Pteropus poliocephalus, Temm. Monog., tom. i. p. 179, tom. ii. p. 66.—Gray, List of Mamm. in Brit. Mus., p. 36.

New South Wales is the true and probably the restricted habitat of this large species of Bat; for I have never seen a specimen from any other part of the Australian continent, and it certainly does not inhabit Van Diemen's Land as stated by M. Temminck : the situations in which I met with it were the dense and luxuriant brushes which fringe the south-eastern portion of Australia, such as those at Illawarra, in the neighbourhood of the Hunter, the Manning and the Clarence; I possess, however, a specimen said to have been killed at Bathurst, which, although of much smaller size, I believe to be the same. Like all other Bats, the Grey-headed Vampire is strictly nocturnal in its habits, and remains during the day suspended from the branches of the larger trees clothing the gullies and mountain sides ; at nightfall it sallies forth in search of its natural food, which principally consists of the fruits and berries peculiar to the brushes, the small wild fig when ripe being a favourite article. The enormous numbers that may be seen sleeping pendent from the trees in the more secluded parts of the forest are beyond conception ; it is not surprising therefore that the settlers whose abodes may be in the neighbourhood of one of these colonies, should find their peach orchards entirely devastated in a single night. Indeed no one of the native animals is more troublesome to the settlers than this large Bat, which, resorting to the fruit-grounds by night, when it is impossible to protect them from its attacks, commits the most fearful havoc. Many pages might doubtless be written respecting the habits and economy of these great Bats, but this can only be done by those who, having been long resident in the country, have had ample opportunities of observing them, which the rapidity of my explorations and the brevity of my stay did not admit. In describing the habits of a nearly allied species (the *Pteropus Javanicus*) Dr. Horsfield states, that " they congregate in companies, and selecting a large tree for their resort, suspend themselves by the claws of their hind limbs to the naked branches, affording to the stranger a very singular spectacle; in short, to a person unaccustomed to their habits, they might be readily mistaken for fruit of a large size, suspended from the branches. They thus pass the greater portion of the day in sleep; but soon after sunset they gradually quit their hold, and pursue their nocturnal flight in quest of food. They direct their course, by an unerring instinct, to the forests, villages and plantations, occasioning incalculable mischief, attacking and devouring indiscriminately every kind of fruit, from the abundant and useful cocoa-nut, which surrounds the dwelling of the meanest peasantry, to the rare and most delicate productions which are cultivated with care by princes and chiefs of distinction. Their flight is slow and steady, pursued in a straight line, and capable of long continuance." This interesting account of the habits of the Javan species doubtless applies in an equal degree to those of the present animal, since we may reasonably infer that the economy of two species so nearly allied is very similar.

Its flesh forms one of the multitudinous articles partaken of as food by the aborigines.

The entire head brown, grisled with grey; round the neck and advancing on to the back a very broad collar of deep rust-red; upper surface and the clothing of the arms glossy black, grisled with greyish olive, the olive hue becoming more apparent on the hind quarters ; under surface brownish black, many of the hairs pointed with olive-yellow; down each flank a patch of rufous; ears and wing-membranes naked and of a deep purplish black ; claws black, becoming horny at the tip.

The figures are of the natural size.

PTEROPUS CONSPICILLATUS.

PTEROPUS CONSPICILLATUS, *Gould.*

Spectacled Vampyre.

Pteropus conspicillatus, Gould in Proc. of Zool. Soc., 1849, p. 109.

The native habitat of this fine species of Vampyre is Fitzroy Island, lying off the eastern coast of Australia, where it was discovered by Mr. John MacGillivray during the recent surveying voyage of H.M.S. Rattlesnake, under the command of the late Capt. Owen Stanley. It is about the same size as the *P. poliocephalus*, but has a somewhat larger head and much larger and more powerful teeth; it may moreover be distinguished from that species by the nuchal band being of a deep sandy buff, instead of deep rust-red, and not continuous round the neck; by the crown of the head and the back being almost jet-black; and by the eyes being conspicuously encircled with deep buff, whence the specific name, and which at once distinguishes it from every other known species.

I am indebted to Mr. MacGillivray for the following brief notes, which comprise all that is at present known respecting it:—

"On the wooded slope of a hill on Fitzroy Island, I one day fell in with this Bat in prodigious numbers, looking while flying along the bright sunshine, so unusual for a nocturnal animal, like a large flock of rooks; on close approach, a strong musky odour became apparent, and a loud incessant chattering was heard; many of the branches were bending under their load of bats, some in a state of inactivity suspended by their hind claws, others scrambling along among the boughs and taking to wing when disturbed. In a very short time I procured as many specimens as I wished, three and four at a shot, for they hung in clusters, but unless killed outright they remained suspended for some time; when wounded they are handled with difficulty, as they bite severely, and on such occasions their cry reminds one of the squalling of a child."

Crown of the head black, slightly grizzled with buff; round each eye a large oval patch of deep brownish buff, which advances on the sides of the face and shows very conspicuously; at the nape a broad crescent-shaped band of deep sandy buff, which extends down the sides of the neck and nearly meets on the breast; centre of the back glossy black, slightly grizzled with grey; cheeks, chin, all the under surface and rump black, slightly grizzled with buff; ears and wing-membranes naked, and of a deep purplish black; claws black.

The Plate represents a male of the natural size.

PTEROPUS FUNEREUS; *Temm.*

PTEROPUS FUNEREUS, *Temm.*

Funereal Vampire.

Pteropus funereus, Temm. Monog. tom. ii. p. 63. tab. 35. fig. 4.—Gray, List of Mamm. in Brit. Mus. p. 37.

Al-wo-re, of the Aborigines at Port Essington.

THIS species appears to be as exclusively confined to the northern portions of Australia as the *Pteropus poliocephalus* is to the south-eastern. M. Temminck gives the animal a very wide range, for he states that he has positive evidence of its existence on four other islands, namely Timor, Amboyna, Borneo and Sumatra. Mr. Gilbert's notes inform me that " it is extremely abundant in all parts of the Cobourg Peninsula; during the day it may be seen in great numbers suspended from the upper branches of the mangroves overhanging the creeks: while living it emits a very strong and disagreeable odour which is perceptible at a considerable distance; it becomes very active at night, and while flying about in search of food utters a loud, trembling, but shrill whistle." Frequent mention is made of this species in Dr. Leichardt's Journal of his Expedition from Moreton Bay to Port Essington, and despite of its disagreeable odour, it often formed for himself and party an excellent and welcome article of food. Like the other species, it feeds upon fruit and the honey of the various flowers; in one instance Dr. Leichardt found them feeding upon the blossoms of the tea-tree, and remarked that they were then more than usually fat and delicate, while those that had been revelling among the blossoms of the gum-trees were not so fat, and had a strong, unpleasant odour. So numerous did they become towards the latter part of the journey, that " twelve were brought in for luncheon, thirty more were procured during the afternoon, and at least fifty were left wounded and hanging to the trees; upon another occasion they were seen clustering in such numbers, that the branches of the low trees drooped with their weight so near the ground, that they could easily be killed with cudgels. In the neighbourhood of the river Roper, myriads were suspended in thick clusters on the highest trees, in the most shady and rather moist parts of the valley; they started as the travellers passed, and the flapping of their large membranous wings produced a sound like that of a hail-storm." Dr. Leichardt went the next day with two of his party to the spot where they had seen the greatest number, and while he was examining the neighbouring trees his companions shot sixty-seven, of which fifty-five were brought to the camp, and served for dinner, breakfast and luncheon, each of the party receiving eight: the animal here lived upon a small, blue, oval stone-fruit, of an acid taste, with a bitter kernel, growing on a tree of moderate size.

Considerable difference is found to exist in the colouring of this animal, but whether this difference is due to sex or age is at present unknown: the following are the descriptions of the two specimens now before me:—

In the one, the head, upper part of the body, the rump, and all the under surface is clothed with a thick, loose black fur, with a wash of deep chestnut between the shoulders; the centre of the back, and the arms, clothed with thin, shining, closely pressed black hairs. In the other, there is a wash of rufous round the eyes, and a broad collar of rich deep chestnut across the nape of the neck. In both the wing-membranes are deep purplish black; and the claws are black.

The figures are of the natural size.

MOLOSSUS AUSTRALIS, *Gray*

J. Gould and H. C. Richter del. et lith.

Hullmandel & Walton, Imp.

MOLOSSUS AUSTRALIS, *Gray.*

Australian Molossus.

Molossus australis, Gray in Mag. of Zool. and Bot., vol. ii. p. 501.—Gould in Proc. of Zool. Soc., July 27, 1858.

THIS large and truly singular species of Bat was described by Dr. Gray just twenty years ago, his description having appeared in the Number of the "Magazine of Zoology and Botany" for December 1838. The specimen from which Dr. Gray took his characters was then and still is the property of the United Service Institution, to whose museum it was presented by Major M'Arthur, with the word "Australia" written on the label attached to it; beyond this nothing was known respecting it, and up to this time it remains the only example in Europe. My thanks are especially due to the President and Council of this Institution for permission to take this rare and valuable specimen out of their cases for the purpose of figuring in the present work; this had been done and my drawing made, when, just on the eve of publication, an unexpected letter arrived from a friend in Australia, of which the following is a copy :—

"Melbourne, Victoria, May 12, 1858.

"DEAR SIR,—A few days ago I saw five Bats together in a hollow tree near this place. Not having seen them figured in your own or any other work, I thought it likely they might prove to be a new species. I therefore made a sketch of one of them for you, and if you think proper to publish it you are quite welcome so to do. I also leave it to you to name the animal. A short description I have the honour to send you.

"Dentition : incisors $\frac{1\cdot1}{2\cdot2}$, canines $\frac{1\cdot1}{1\cdot1}$, false molars $\frac{2\cdot2}{2\cdot2}$, true molars $\frac{3\cdot3}{3\cdot3}=\frac{14}{16}=30$.

"For the admeasurements of the animal I refer you to the sketch, which is of the size of life. Colour of the body sepia-brown, the belly somewhat lighter; wings greyish brown; hand or hind-foot and the extremities, including the tail, black; the under side of the fore-arms whitish flesh-colour; the palm and wrist of the hand black, as if covered with gloves; between the elbows and the knees a pure white streak stretching towards the root of the tail; irregular white spots occur on the neck and chest in some of the specimens, in others the neck and belly are covered with large white patches; tragus and the ears, where free from hairs, black, the remainder clothed with dark rust-coloured hairs; on the upper margin of the ear a row of diminutive tubercles; eyes black; space surrounding the nostrils naked and black; under lip nude and of a blackish brown; thumb and fourth finger of the hind-foot thicker than the three middle ones, while a sort of fine brush covers the former; the thumb is shorter than the fingers, but all have on the top of the nail a small tuft of fine hairs; the tail is prolonged for more than an inch beyond the intra-femoral membrane; on the throat a sort of pouch stretching inwards and downwards nearly three-quarters of an inch, covered with two distinct tufts of stiff brown hairs growing on the bottom of the pouch, and resembling a couple of artist's brushes; when the pouch is not distended, only the extremity of the brushes are visible, and are scarcely distinguishable from the other hairs; the naked portion of the pouch is flesh-coloured; there is no communication between the pouch and the inner portion of the neck. The upper incisor teeth are rather large, and resemble canines; the lower ones are very minute; the upper first and second true molars are nearly equal, and have three sharp tubercles externally; the third is smaller, and has two pointed tubercles; the first and second of the lower true molars have five points on their crowns.

"I shall be very much pleased to receive a few lines stating your opinion of the Bat, and if I can serve you in any way connected with natural history.

"With all due respect,

"Yours obediently,

"LUDWIG BECKER."

The drawing which accompanied this letter accorded so nearly with the specimen in the United Service Museum as not to leave a shadow of doubt as to the animals discovered by Dr. Becker being identical with it; the tufts of hair and the pouch are, however, almost obsolete in the specimen, which is probably due to a difference in sex, or of the season at which it was killed.

On submitting Dr. Becker's letter and drawing to R. F. Tomes, Esq., who has paid great attention to the *Vespertilionidæ*, that gentleman favoured me with the following remarks :—

"Welford, Stratford-on-Avon, June 1858.

"MY DEAR SIR,—I have compared the drawing and description of the *Molossus Australis* with a great many species of that genus in spirit and prepared skeletons, and conclude most certainly that it is a

Molossus. The only doubt I had was whether the genus *Nyctinomus* of Isidore Geoffroy should properly be separated from the genus *Molossus* established by his father. In order to determine this, if possible, I have cleaned a number of specimens and instituted a minute comparison. The results scarcely justify a *sub-generic* difference. With respect to the gular pouch, that appears, so far as my specimens in spirit inform me, to be peculiar to the Australian species; but I strongly suspect that it will be found to be a peculiarity due to sex and age, perhaps even in some measure to season, rather than of generic value. In the genus *Taphozous* this gular sac is entirely dependent on the age and sex of the animal, and its absence or presence has been made use of as a specific character very improperly.

"It is not a little remarkable that this Australian species should possess characters (with the exception of the pouch) which are as similar to the European ones as to any other species. I have often been surprised that Australia does not furnish a single form among the Bats that are not common to nearly all the world besides ; indeed, many of the species are found in the Indian islands, and, curiously enough, in *China*. A collection of Chinese Bats which I have lately examined consisted of Indian and Australian species. I have taken very great pains in the examination of the drawing, and have prepared skeletons of several species on purpose to compare them. I would strongly advise you to continue the name of *Molossus Australis*, and you may, if you like, add that the animal belongs to that division of the genus which has been designated *Nyctinomus.*   " Ever yours truly,

"R. F. TOMES."

This, then, is all that I am at present able to communicate respecting the species, and it affords me much pleasure to furnish even this meagre account, for a more interesting animal I have not seen for some time.

The following is a description of the animal in the United Service Museum :—

Fur soft and dense, that of the upper surface of a deep reddish brown, and extending on to the basal parts of the limbs ; fur of the under surface of a similar but much lighter hue, bordered on each side of the body by a broad fringe of white, which extends in a lesser degree on to the base of the lower limbs and the posterior part of the body ; wing-membranes extremely thin, and presenting a silvery appearance ; ears and face dark purplish brown.

Total length 6⅜ inches ; extent of the wings from tip to tip 15¼.

The figure is of the natural size.

TAPHOZOUS AUSTRALIS, *Gould*

J.Gould and H.C.Richter, del. et lith.

Hullmandel & Walton, Imp.

TAPHOZOUS AUSTRALIS, *Gould.*

Australian Taphozous.

For a knowledge of this fine species of Bat, we are indebted to the researches of Mr. John MacGillivray, the Naturalist attached to H.M.S. Rattlesnake, during the recent survey of the northern coasts of Australia. After carefully comparing it with the various species contained in the unrivalled collcetion of the Vespertilionidæ in the Museum at Leyden, I have come to the conclusion that it is quite distinct, and have, therefore, given it the name of *Taphozous Australis,* it being, I believe, the first species of the form found in that country.

Mr. MacGillivray having kindly furnished me with a copy of his notes respecting this species, I cannot, perhaps, do better than give them in his own words ;—notes taken on the spot being of infinitely greater value than any that can be elaborated from the most careful examination of the dried specimen.

" Dentition :—incisors $\frac{1 \cdot 1}{2 \cdot 2}$; canines $\frac{1 \cdot 1}{1 \cdot 1}$; false molars $\frac{2 \cdot 2}{2 \cdot 2}$; true molars $\frac{3 \cdot 3}{3 \cdot 3} = \frac{14}{16} = 30.$

" Length 3 inches ; tail, 1 ; fore-arm, 2·5 ; hind-arm, 1 ; ear, 1 ; tragus, 0·25.

" Colour (there are two varieties) : above, ferruginous brown ; light brown in the centre of the back and across the abdomen, or entirely brownish grey ; basal half of the fur white ; below, ash-grey, with sometimes a slight reddish tinge ; muzzle black.

" Nostrils : simple, terminal, bordering the upper lip.

" Ears : not connected at the base ; rather large ; somewhat triangular, the two exterior angles rounded ; within thinly covered with scattered hairs ; upper margin with a row of small tubercles ; tragus one-fourth the length of the ear, irregularly quadrate ; narrowed at the base ; the exterior angles rounded, the inferior projecting most.

" Wings : above with a few hairs at the elbow, and rather thickly covered at the base of the intra-femoral membrane ; below with a band of scattered whitish hairs along the arm and fore-arm as far as the fifth digital phalange ; second phalange single-jointed ; the third, fourth and fifth three-jointed.

" Tail not reaching to the posterior margin of the iutra-femoral membrane ; free for two-tenths of an inch.

" Incisors : upper very minute, simple ; lower small, three-lobed.

" Canines : upper strong, curved, acute, with a small basal lobe projecting behind, and internally forming a sharp point ; lower more slender, with projecting collar in front and on the sides.

" False molars : above, first very minute, second long, acute : below, acute, the second rather the longest.

" True molars : above, first and second nearly equal, with two sharp tubercles externally, and a low, sharp ridge internally ; third very small, transverse, with two pointed tubercles and a sharp ridge ; below, the first and second with five, and the third, which is the smallest, with four points on the crown.

" Habitat : the maritime caves in the sandstone cliffs of Albany Island, Cape York. In great numbers in three of the caves. Specimens obtained October 1848."

The animal is represented of the natural size.

RHINOLOPHUS MEGAPHYLLUS, Gray

J.Gould and H.C.Richter, del.et lith.

Hullmandel & Walton, Imp.

RHINOLOPHUS MEGAPHYLLUS, *Gray.*

Great-leaved Horse-shoe Bat.

Rhinolophus megaphyllus, Gray in Proc. of Zool. Soc., part ii. p. 52.—Ib. List of Mamm. in Coll. Brit. Mus., p. 22.

THIS species, the largest of the Horse-shoe Bats that has yet been received from Australia, was described by Dr. Gray as long back as 1834, from a specimen collected in the caverns in the neighbourhood of the river Morumbidgee, in New South Wales. The example from which my figure was taken was obtained by Mr. Leycester near the Richmond river. I mention these localities particularly, because the animal is one of those which did not come under my notice during my explorations in Australia, although it was almost a never-failing practice with me to devote the last hour of the day to the study and collection of these curious little mammals.

Dr. Gray states, that "this Bat is very nearly allied to the true European *Rhinolophi*, and agrees with them in having four cells at the base of the hinder nose-leaf, and distant pectoral teats, but differs from them in having a much broader nose-leaf.

"The hinder nose-leaf is bristly, ovate-lanceolate, nearly as broad at the base as the face, with a rather produced tip; the septum of the nose is grooved; and the front leaf expanded with a quite free membranaceous edge. The head is elongated; the face depressed; the muzzle rounded; the ears are large, reaching when bent down rather beyond the tip of the nose. The fur is soft and of a pale mouse-colour. The membranes are dark and naked, with rather distant whitish hair on the under side, near the sides of the body."

The figures are of the size of life.

RHINOLOPHUS CERVINUS, *Gould*

J.Gould and H.C.Richter del et lith.   Hullmandel & Walton, Imp.

RHINOLOPHUS? CERVINUS, *Gould.*

Fawn-coloured Bat.

I HAVE figured this species as a *Rhinolophus* with a mark of doubt, being somewhat uncertain as to whether I am correct in placing it in that genus; probably it ought to have been assigned to that of *Phyllorhina.* Mr. MacGillivray, to whom we are indebted for its discovery, was inclined to think it identical with *Rhinolophus aurantius,* but upon comparing it with that species, I am convinced it is distinct; I have therefore assigned it a specific appellation, and have selected that of *cervinus,* in reference to the colouring of the fur. The following notes respecting the animal were communicated to me by Mr. MacGillivray, and as they were made at the time he procured the specimens from which my figures are taken, it will be well perhaps to give them in his own words :—

"Dentition : incisors $\frac{1 \cdot 1}{2 \cdot 2}$; canines $\frac{1 \cdot 1}{1 \cdot 1}$; false molars $\frac{2 \cdot 2}{2 \cdot 2}$; true molars $\frac{3 \cdot 3}{3 \cdot 3} = \frac{14}{16} = 30.$

"Length: body, exclusive of the tail, 2 inches; fore-arm, 1·7; hind-arm, 0·7; tail, 1; ears, 0·5 long, 0·45 wide; extent of wings, 11 inches.

"Colour: above tawny brown, darkest on the face, head and shoulders; below paler, and tinged on the belly with grey.

"Nose-leaf simple, long, straight-edged, 0·25 across.

"Ears : connected by a hairy fold of skin, large, broadly ovate, pointed; posterior margin slightly sinuated near the tip, then rounded; internally with anterior one-third thickly clothed with hair; tragus obsolete, being indicated merely by a slight internal fold of the auricle.

"Wings naked; index one-jointed, the others three-jointed.

"Tail continued 0·1 beyond the intra-femoral membrane.

"Incisors : above very minute; below larger and three-lobed.

"Canines : strong, hooked, sharp, the upper ones the largest.

"False molars : above, first very minute, second large and pointed; below, simple, pointed, the second the largest.

"True molars : first and second in each jaw with five, and the third with four sharp points.

"Habitat : Cape York; also in the sandstone caves on Albany Island, where it occurs in great numbers. The two species do not associate together. Procured October 1848."

The figures are of the natural size.

RHINOLOPHUS AURANTIUS.

J. Gould and H.C. Richter, del. et lith.   Hullmandel & Walton, Imp.

RHINOLOPHUS AURANTIUS, *Gray.*

Orange Horse-shoe Bat.

The Orange Horse-shoe Bat (Rhinolophus aurantius), Gray, App. to Eyre's Journ. of Exp. of Disc. into Central
Australia, vol. i. p. 405. tab. 1. fig. 1.

———————

THE only information we possess respecting this beautiful Bat, is that it is abundant on the Cobourg Peninsula in Northern Australia; that it retires during the day-time to the hollow spouts and boles of the
various species of *Eucalypti*; and that it sallies forth on the approach of evening in search of its insect
food : its general habits and manners in fact so closely resemble those of the other members of the genus,
that a separate description of them is quite unnecessary.

Mr. Gray, who characterized the animal in the Appendix to Mr. Eyre's "Travels," above referred to, from
a specimen procured while flying near the Hospital at Port Essington, by Dr. Sibbald, R.N., remarks that it
is "peculiar for the brightness and beauty of its colour, the male being nearly as bright as the Cock of the
Rock (*Rupicola aurantia*) of South America."

The following is Mr. Gray's description of this pretty animal :—

"Ears moderate, naked, rather pointed at the end; nose-leaf large, central process small, scarcely lobed,
blunt at the tip; fur elongate, soft, bright orange; the hairs of the back with short brown tips, of the under
side rather paler, of the face rather darker; membranes brown, nakedish; tail rather produced beyond the
membrane at the tip; feet small and quite free from the wings.

"The female pale yellow, with brown tips to the hair of the upper parts."

The figures are of the natural size.

NYCTOPHILUS GEOFFROYI, *Leach* *

J. Gould and H.C. Richter del et lith.   Hullmandel & Walton, Imp.

NYCTOPHILUS GEOFFROYI, *Leach**.

Geoffroy's Nyctophilus.

Nyctophilus Geoffroyi, Leach in Linn. Trans., vol. xiii. p. 73, 1820–22.—Less. Man. de Mamm., p. 86, 1827.—
Fisch. Synops. Mamm., p. 135, 1829.—Temm Mon., tom. ii. p. 47, 1835–41.—Wagn. Supp. Schreib.
Säugeth., tom. ii. p. 442, 1840.—Less. Nouv. Tab. Règn. Anim., p. 33, 1842.—Schinz, Synops. Mamm.,
tom. i. p. 217, 1844.—Tomes in Proc. of Zool. Soc., part xxvi. p. 29.

SINCE my plate and description of the animal I have called *Nyctophilus Geoffroyi* were printed, Mr. Tomes has very minutely investigated this group of bats, and published a monograph of the genus, and he now considers that the hint I there gave as to the probability of the species from Western Australia and Tasmania being distinct is a correct view of the case, and has come to the conclusion that the Western Australian species is the true *N. Geoffroyi*, and consequently that the animal from New South Wales, formerly figured by me under that name, should receive a new appellation; and he has accordingly named it after myself, *N. Gouldi*. It is much to be regretted that this conclusion should not have been arrived at before my plate and description were printed, as the synonymy of the New South Wales *N. Gouldi* has reference to the animal here represented, which is a native of Western Australia; however as Mr. Tomes's opinions are of value, and entitled to be recorded in this or any other work comprising an account of any of the members of the family *Vespertilionidæ*, I will quote his own words:—

"This, from its size," says Mr. Tomes, "is unquestionably the species on which Dr. Leach established the genus. The original description in the Linnean Transactions is much too vague to discriminate the exact species with certainty; but M. Temminck having become possessed of the original specimen, and given a more detailed description of it, I am enabled to determine with certainty which is the true *N. Geoffroyi*."

Mr. Gilbert states that this species is called *Bar-ba-lon* by the aborigines of King George's Sound, and *Bambe* by the natives of Perth, and that it is the most abundant species in the colony of Western Australia. It is sometimes met with by the wood-cutters in the hollow spouts of the gum-trees in great numbers; from these retreats they emerge at twilight and flit about the shrubs and lower trees in search of insects.

The following is Mr. Tomes's description:—

"The face is moderately hairy, the hairs being pretty regularly scattered, but a little thicker on the upper lips and on the second nose-leaf than elsewhere; immediately over the eye is a small tuft of bristle-like black hairs, and a similar one near the hinder corner of the eye; at the angle of the mouth a few similar hairs may be observed; the fur of the back extends to a very trifling extent on to the interfemoral membrane, but all the other membranes are perfectly naked and of a dark brown colour, as are also all the other naked parts, with the exception of the tragus and the contiguous parts of the inside of the ear, which are brownish yellow. The fur of the body is rather long, thick, and very soft; on all the upper parts it is conspicuously bicoloured, black for nearly two-thirds of its length, the remainder being olive brown, of which the extreme tips are rather the darker portion; on the membranes uniting the ears the fur is uniform yellowish brown; the fur of the throat and flanks is uniform brownish white, that of the latter being sometimes more strongly tinted with brown; all the remaining underparts have the fur markedly bicoloured black at the base, with the terminal third brownish white, varying considerably in purity of colour in different individuals."

"This description," says Mr. Tomes, "was taken from a specimen kindly lent to me by Mr. Gould, and which is labelled ' Albany, King George's Sound, May 19, 1843.' "

The figures are of the natural size.

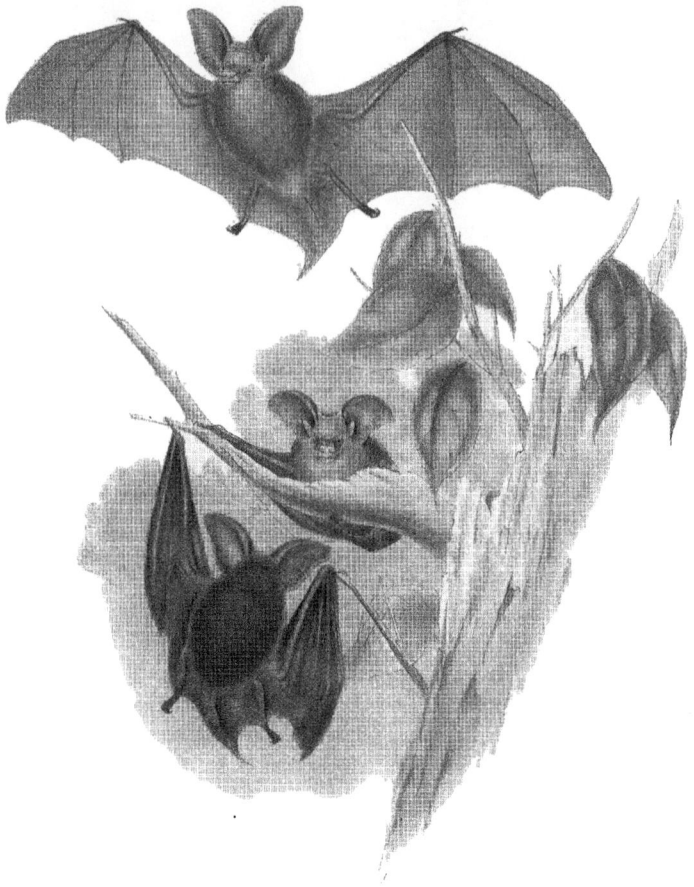

NYCTOPHILUS GEOFFROYI, Leach

Gould and H.C Richter, del et lith.   Hullmandel & Walton, Imp.

NYCTOPHILUS GEOFFROYI.

Geoffroy's Nyctophilus.

Nyctophilus Geoffroyi, Leach, in Linn. Trans., vol. xiii. p. 78.—Temm. Monog., vol. ii. pl. 34.—Gray, in Mag. of Zool. and Bot., vol. ii. p. 12.—*Id.*, List of Mamm. in Coll. Brit. Mus., p. 25.—*Id.*, in Grey's Journ. of Two Exp. in N.W. and W. Australia, vol. ii. p. 400.

Barbastellus Pacificus, Gray, Zool. Misc., vol. i. p. 38.

Nyctinomus ——, Benn. Cat. of Australian Museum, Sydney, p. 1. no. 2.

THE figures on the accompanying Plate are taken from specimens captured in New South Wales, and I consider it necessary to state this particularly, because the long-eared Bat of Western Australia, though very nearly allied, may prove to be distinct : the specimens I possess from the latter country are larger, and have much more powerful teeth than any examples I have seen from the eastern parts of the continent. I shall therefore speak of the present animal as an inhabitant of New South Wales and Van Diemen's Land, with a slight doubt as to whether the Tasmanian animal may not be different also, all the specimens I have yet examined being smaller and darker than those from New South Wales or Western Australia. Every mammalogist is aware how closely the *Vespertilionidæ* are allied, and how difficult it is to obtain correct information respecting the species inhabiting our own country. I may therefore be readily excused for not coming to a hasty conclusion on the subject of those of the antipodes : one thing is certain, namely, that the animal figured is identical with the specimens in the British Museum which were received from New South Wales, and to which I find the name of *Geoffroyi* attached.

I frequently saw this animal during my sojourn in New South Wales, and remarked that it was a high flier and extremely active in the air ; in other respects, as may be supposed, it closely assimilated in its actions and economy to the nearly allied species in Europe. As the figures in the accompanying Plate are the size of life, it will be unnecessary to give the admeasurements.

The face is fleshy brown, deepening into dark brown on the nose and laterally expanded nose-leaf ; fur clothing the upper surface brown, that of the under surface greyish brown, washed with rufous on the sides ; ears and wing-membranes purplish brown.

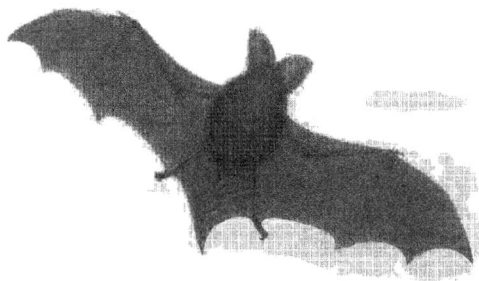

NYCTOPHILUS UNICOLOR, *Tomes*

J. Gould and H. C. Richter, del. et lith.   Hullmandel & Walton, Imp.

NYCTOPHILUS UNICOLOR, *Tomes.*

Tasmanian Nyctophilus.

Nyctophilus unicolor, Tomes in Proc. of Zool. Soc., part xxvi. p. 33.

———————

" ALL the specimens of this genus I have yet seen from Van Diemen's Land," says Mr. Tomes, " differ remarkably from those of the mainland of Australia, in having the fur everywhere short and cottony, perfectly devoid of lustre, and unicoloured; that of the upper parts is of a dark olive-brown without any variation of tint, excepting that it is perhaps a little darker along the middle of the back than elsewhere; beneath the fur is similar but paler in colour, with the tips of the hairs a little tinged with ash-colour; this is the colour of the whole of the under parts, with the exception of a patch on the throat, which is whitish brown, dirty white, and occasionally pure white.

"Immature examples often have the fur above and beneath of a very dark olive-brown, almost black. One specimen of this dark colour which I have examined has the spot on the throat almost pure white.

" So far as I have been able to ascertain, this species is subject to very trifling variations, either in colour or size in the adult state; and the size agrees so closely with that of the species which I have called *N. Gouldi*, that I at first thought the great difference in the texture and colour of the fur was due to the difference of locality."

To this description I have nothing to add. The specimens in my collection were transmitted from Tasmania to this country by Ronald C. Gunn, Esq., a gentleman who has done much to enrich our knowledge of natural history.

The upper figure is of the natural size, the lower one somewhat reduced.

NYCTOPHILUS TIMORIENSIS.

Western Nyctophilus.

Vespertilio Timoriensis, Geoff. Ann. du Mus., tom. viii. p. 200. tab. 47, 1806.—Desm. Mamm., p. 146, 1820.—
  Fisch. Synop. Mamm., p. 118, 1829.—Temm. Mon., tom. ii. p. 253, 1835–41.—Wagn. Supp. Schreib.
  Saugth., tom. i. p. 520, 1840.—Schinz, Synop. Mamm., tom. i. p. 175, 1844.
———— ————?, Temm. Mus. Leyd.
Plecotus Timoriensis, Less. Man. de Mamm., p. 97, 1827.—Is. Geoff. in Guérin, Mag. de Zool. 1832.—Less. Nouv.
  Tab. Règn. Anim., p. 23, 1842.
Nyctophilus Timoriensis, Tomes in Proc. of Zool. Soc., part xxvi. p. 30.
Bam-ba, Aborigines of Perth in Western Australia.

I⊤ is believed by Mr. Tomes that this species of Bat, although bearing the name of *Timoriensis*, is never found in Timor, but that its true habitat is Western Australia; certain it is that it was there found by Mr. Gilbert, who states that it is very abundant in the neighbourhood of Perth, that it often flies into the houses, doubtless attracted by the light, and that its flight is extremely rapid.

" Although the original specimen of this species," says Mr. Tomes, " is reported to have been received from Timor, I am inclined to believe that there may have been some mistake respecting its locality. Among a great number of Bats from that island, contained in our museums and in that of Leyden, representatives of this genus do not appear ; but specimens absolutely identical with the original in the Paris Collection have been obtained by Mr. Gould from Western Australia, and I have noted one in the Leyden Museum also from Australia, but without any precise indication of locality.

" The forms of this species are so similar to those of *N. Geoffroyi*, that it is needless to enter at greater length into details of description than is necessary to point out the differences between the two.

" In all the specimens I have been able to examine, viz. the original one in the Paris Museum, and three others collected in Australia by Mr. Gould, the ears are strongly sulcated, even more so than is observable in the *Plecotus auritus*, whilst in the *N. Geoffroyi* they are very faintly if at all marked ; and instead of the small tufts of bristle-like hairs about the eyes, the present species has a tolerably regular series of similar ones fringing the eyelids.

" But the great difference in the size of the two animals is alone sufficient to distinguish them, the one being only nine inches in expanse, whilst the other attains fully thirteen inches ; nearly as great a difference as exists between the *Pipistrelle* and the *Noctule* Bats.

" The fur of the upper parts is bicoloured, nearly black at the base, with the terminal half dark sepia-brown ; that on the top of the head and on the membrane uniting the ears, unicoloured and paler ; beneath, the fur has the basal half nearly black, the remainder being light brown, palest on the throat, on the middle of the belly, and on the pubes ; on the shoulder of one example from Perth, Western Australia, is a patch of brownish rust-colour, but it does not occur in the other examples.

" This animal has been repeatedly described as a *Vespertilio*—*V. Timoriensis* ; but it is strictly a *Nycto-philus*, as I have ascertained by the examination of the original specimen in the Paris Museum."

The figure is of the size of life.

SCOTOPHILUS GOULDI, Gray.

J Gould and H.C. Richter, del et lith

Hullmandel & Walton, Imp

SCOTOPHILUS GOULDI, *Gray.*

Gould's Bat.

Scotophilus Gouldii, Gray in Grey's Journ. of Discoveries in Australia, App. vol. ii. p. 405.—Ib. List of Mamm. in Coll. Brit. Mus., p. 30.

This fine species of Bat is very generally dispersed over New South Wales, and, I believe, South Australia; but, as yet, I have only seen examples from the districts of the former country lying between the mountain ranges and the sea, where it frequents the outskirts of the brushes and the wooded borders of the great rivers. It may be readily distinguished by the upper half of the body being black, while the lower is suffused with brown; and by the hairs of the latter hue on the under surface being lengthened, and extending on to the arms and wing-membranes. It appears, however, to be subject to considerable variation in colour, some being parti-coloured as described, while in others the black predominates; others again, from Flinders' Range in South Australia, have the brown tint reaching nearly to the nape on the upper surface and to the chest on the under surface. I have some specimens also from this locality with a good deal of brown on the chin and throat. I was for some time inclined to consider the Flinders' Range specimens to be distinct; but, on submitting them to the inspection of Mr. Tomes, who has paid the most minute attention to this group of animals, that gentleman states that he considers them to be identical, and that the mere variation in colour, unaccompanied by a difference in structure, is not sufficient to warrant their separation.

The anterior half of the body, both above and beneath, is sooty-black; the posterior half of the upper surface brown; sides and abdomen brownish fawn-colour; wing-membranes purplish-brown.

The figures are of the natural size.

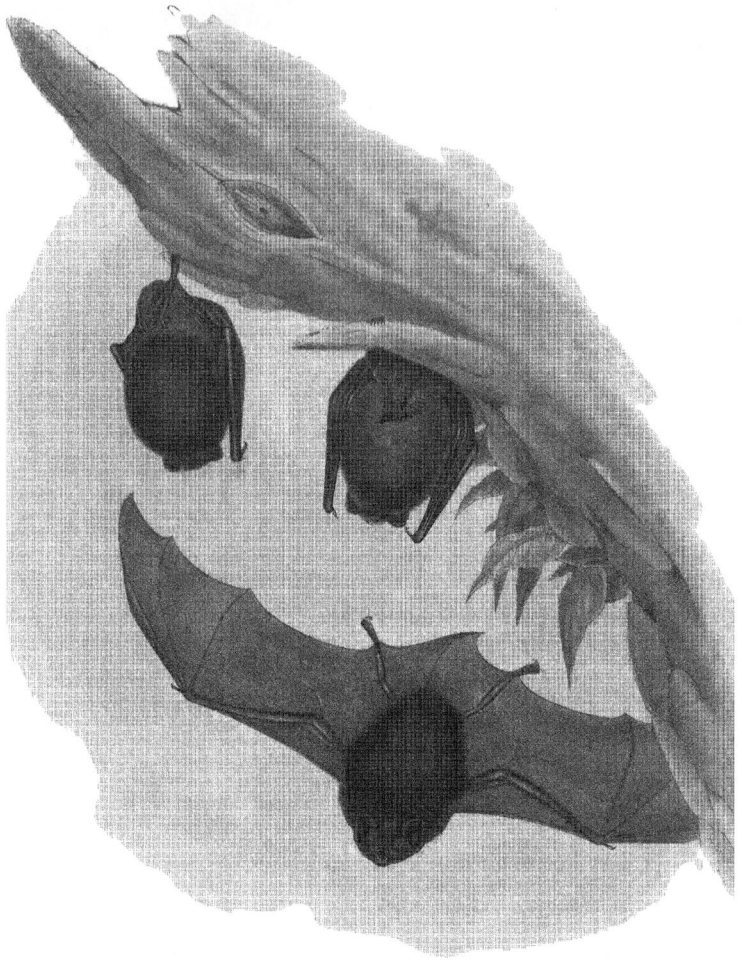

SCOTOPHILUS MORIO, Gray

J. Gould and H C Richter del et lith

Hullmandel & Walton, Imp

SCOTOPHILUS MORIO, *Gray*.

Chocolate Bat.

Scotophilus morio, Gray in Grey's Journ. of Discoveries in Australia, App. vol. ii. p. 405.—Ib. List of Mamm. in Coll. Brit. Mus., p. 29.

THIS species is about the size of *Scotophilus Gouldi*, but differs in having larger ears, and in the colouring of the entire body being of a uniform chocolate-brown. It is very common in New South Wales, between Moreton Bay and Sydney, and Mr. Gilbert states that it also inhabits Western Australia. I have not, however, his specimens to compare with those from New South Wales; its inhabiting the western coast must therefore rest upon his authority; if his assertion be correct, its range will probably be found to extend over the whole of the southern portion of the country. The animal Mr. Gilbert describes is called by the natives *Bam-be*, and in his notes he says that "it is rather uncommon, but may be readily recognized by its habit of flying at a great elevation, and generally around the branches of the loftiest *Eucalypti*."

The whole of the fur of both the upper and under surface of a uniform chocolate-brown, becoming somewhat darker or nearly black on the cheeks; wing-membranes purplish-brown.

The figures are of the natural size.

SCOTOPHILUS MICRODON, *Tomes*

J.Gould and H.C.Richter del. et lith.   Hullmandel & Walton, Imp

SCOTOPHILUS MICRODON, *Tomes.*

Small-toothed Bat.

Scotophilus microdon, Tomes in Proc. of Zool. Soc., part xxvii. p. 68.—Ann. and Mag. Nat. Hist., 3rd ser. vol. v.
    p. 50.

Mr. Tomes has very kindly favoured me with the loan of a specimen of the Bat represented in the accompanying Plate, for the purpose of enriching the 'Mammals of Australia.' This gentleman, believing the species to be entirely new to science, has characterized it in the ' Proceedings of the Zoological Society of London' for the year 1859, and I cannot perhaps do better than reproduce Mr. Tomes's account of the species, a course which I feel assured will be approved of by every mammalogist, from the confidence we all place in the investigations made by that gentleman.

"The present species is one having the same subgeneric characters as the common *Pipistrelle* of Europe and the *Scot. Greyii* and *S. pumilus* of Australia. To the latter species it is, by the form of its head and ears, most nearly affined, but may at once be distinguished from it by its greater size and by its smaller teeth.

"The crown is but little elevated above the facial line; but the muzzle, although short, is more pointed than is usual in the flat-crowned species. The ears are very small, nearly as broad as high, with the outer margin slightly hollowed out about the middle, below which is a faintly developed lobe, and immediately above which is the tip of the ear,—the latter being obtusely angular, and directed outwards. The inner margin is very much rounded, especially at two-thirds of the distance from the base, where the convexity is so prominent as to be quite as high as the tip itself, the portion between this prominence and the tip being nearly horizontal. Altogether the ear bears some resemblance to that of *Miniopteris*. *Scot. pumilus* is the only species which has ears of form similar to those of the present species; but they are, although the species is smaller, rather larger, relatively longer, and have their tips less outwardly directed and more rounded. The tragus, as in all others of this group, is curved inwards, and rounded at the end; but it differs from that of some others in being rather widest in the middle.

"In relation to the size of the animal, the wings are rather ample, and rather broad for their length, the fourth finger (that which determines the breadth of the wing) being longer than the two basal phalanges of the longest finger. All the wing-bones are somewhat slender. The thumb is rather long, not quite half enveloped in the membrane.

"The legs are rather long and slender, the tibiæ being quite as long as in *S. Gouldii*, a species of greater size than the present; they are just twice the length of those of *S. pumilus*. The feet are large, about the length of those of *S. Leisleri* of Europe, the toes taking up half their entire length, and the wing-membranes extending to half the distance between the extremity of the tibia and the base of the toes. Tip of the tail enclosed in the membrane.

"The fur of the head extends to rather near the end of the nose; and the upper lips are furnished with moustaches; so that the only naked space is around and in front of the eye. The fur of the back does not extend on to the interfemoral membrane, and only to a very limited extent on those of the wings; but that of the under parts encroaches on the membranes all round the body, especially beneath the arms, where it reaches nearly to the elbow. A straight line from that joint to the knee would pretty accurately define the hairy portions of the wing-membranes.

"In quality the fur is soft, and rather long, bicoloured above and beneath. That of the back of a specimen from South Australia is dark brown at the root, with the terminal half of the hairs reddish brown, uniformly of the latter colour around the rump and on the flanks; beneath, dark brown at the root, with the terminal third light cinnamon-brown, that on the membranes paler and unicoloured. Membranes lightish brown.

"Another specimen from Van Diemen's Land differs only from the last in being much darker in colour; the fur of the upper parts black at the root, tipped with sepia-brown; beneath, the same, but the brown tips lighter and more tinged with rufous, especially that on the membranes and around the pubal region, where it is unicoloured and reddish brown.

"The teeth of this species, although not sufficiently examined to furnish a comparative description, are nevertheless seen at a glance to be of very small size, not only in reference to the size of the animal, but also actually smaller than those of several other species of much less size, such as *S. trilatitius*, *S. lobatus*, and *S. abramis*; hence the specific name of *microdon* here bestowed upon it."

The figure is of the natural size.

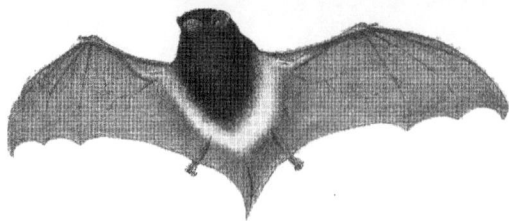

SCOTOPHILUS PICATUS, *Gould.*

Gould & Richter, del et lith.
Hullmandel & Walton, Imp.

SCOTOPHILUS PICATUS, *Gould.*

Pied Scotophilus.

Vespertilio—-Little Black Bat, Sturt, Exp. into Central Australia, vol. ii. App., p. 9.

Scotophilus picatus, Gould in Proc. of Zool. Soc., 1852.

———————

THIS pretty little Bat, which is the smallest and one of the most interesting of the true *Scotophili* inhabiting Australia, is so rare, that the single specimen, procured by my friend Captain Sturt, during his late hazardous journey towards the centre of that country, is the only one that has come under my notice; and all the information at present known respecting it is contained in the following note, given in the Appendix to the second volume of Captain Sturt's valuable account of his expedition quoted above:—

"This diminutive little animal flew into my tent at the depôt, attracted by the light. It is not common in that locality, or any other that we visited. It was of a deep black in colour, and had smaller ears than usual."

The whole of the fur both of the upper and under surface deep glossy black, with the exception of a crescentic mark of white which bounds the sides and the lower part of the abdomen; wing and tail membranes purplish brown.

The figures are of the natural size.

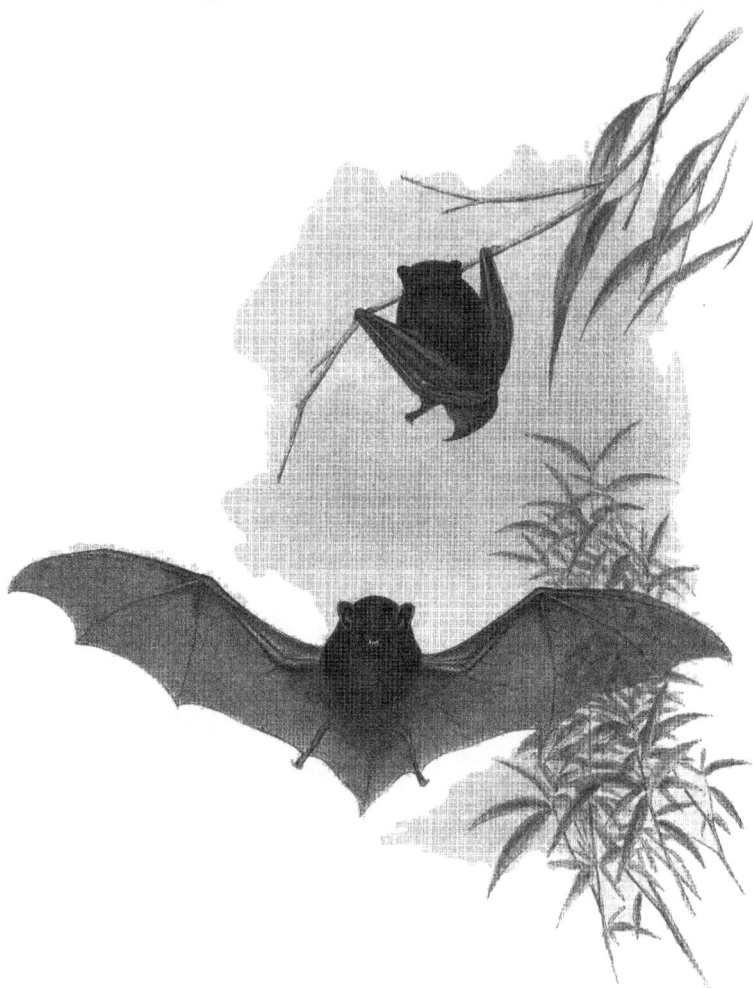

SCOTOPHILUS NIGROGRISEUS, *Gould.*

J. Gould and H.C. Richter, del. et lith.   Hullmandel & Walton Imp.

SCOTOPHILUS NIGROGRISEUS, *Gould.*

Blackish-grey Scotophilus.

A VERY fine specimen of this Bat was sent to me by Mr. Strange, who collected it in the neighbourhood of Moreton Bay. In size it is about equal to the *Scotophilus picatus*, to which it bears a close resemblance, but from which it is quite distinct. The *S. picatus* is an inhabitant of the distant interior, where it was collected by Captain Sturt in the neighbourhood of his farthest encampment, when he endeavoured to reach the centre of the continent from South Australia; the present animal, on the other hand, inhabits the country near the coast; it will be seen therefore that the two species affect very different localities.

The specimen from which my figure was taken will hereafter be deposited in the British Museum, where it may be examined by any mammalogist who may be desirous of investigating the singular group of animals to which it pertains. I may add, that Mr. Tomes, who has paid much attention to this group, coincides with me in considering it to be a new and distinct species from any previously described.

Fur soft and velvety to the touch, the general hue greyish-black, becoming somewhat paler on the posterior part of the upper surface; abdomen washed with brown, and fading into very light brown on the vent; wing- and tail-membranes purplish-brown.

The figures are of the natural size.

SCOTOPHILUS GREYI, Gray

J. Gould and H. C. Richter del. et lith.   Hullmandel & Walton, Imp.

SCOTOPHILUS GREYI, *Gray*.

Grey's Scotophilus.

Scotophilus Greyii, Gray, List. of Mamm. in Coll. Brit. Mus., p. 30.—Ib. Zool. of Voy. of Erebus and Terror,
pl. 20, fig. 2.

This diminutive species was named by Dr. Gray in his " List of the Specimens of Mammalia in the Collection of the British Museum," and was also figured, as above stated, in the " Voyage of the Erebus and Terror." It is said to be a very rare species, and not to have been hitherto found in any other part of Australia than Port Essington. Dr. Gray has given it the above appellation in honour of Sir George Grey, the present Governor of the Cape of Good Hope, to whom such a tribute is justly due for his devotion to the natural sciences generally: from his enlightened views much good has already accrued to every community over which Sir George has had influence, and, as a traveller, he must be considered one of the most intrepid of England's sons.

The fur of this little animal is of a light reddish brown, somewhat paler on the under than on the upper surface; the nose is reddish flesh colour, and the wing-membranes of the usual purplish brown, as seen in other members of the genus.

The figures are of the natural size.

SCOTOPHILUS PUMILUS, *Gray*.

Little Bat.

Scotophilus pumilus, Gray in App. to Grey's Two Exp. in N.W. Australia, vol. ii. p. 406.—Ib. List of Mamm. in Brit. Mus., p. 30.

ALTHOUGH the *Vespertilionidæ* are fairly represented in Australia, the species inhabiting that country are not very numerous. The Bat here represented is certainly one of the very least of the Australian members of the family, for it scarcely exceeds in size the European Pipistrelle. It was my usual practice when travelling in Australia to look around me during the last half-hour of daylight for Bats, at which to discharge the contents of my gun before retiring to my tent, and by this means several species were collected, which might otherwise even now be unknown in Europe. It was not, however, always necessary to shoot this little animal, for it is very tame, and my black attendants often amused themselves by cutting it down with a switch as it passed before them, or rapidly skimmed over the water, a frequent habit with it. I found it especially abundant on the upper part of the River Hunter, particularly on the banks of the rivulets descending from the mountain ranges.

I have never heard of the *Scotophilus pumilus* being collected by any one but myself, and I regret to say that I am unable to give any details as to its habits and economy.

Fur of the upper surface greyish-brown, and of a darker or blackish hue at the base; under surface paler; cheeks blackish; wing-membranes purplish-brown.

The figures are of the natural size.

VESPERTILIO MACROPUS, *Gould.*

Great-footed Bat.

Mr. Tomes having carefully examined my collection of Bats, and come to the conclusion that this animal has not been described, I have, in accordance with his views, characterized it as distinct. It is a native of South Australia, in every respect a true *Vespertilio*, and remarkable for having rather lengthened and elegantly-formed ears, a delicately-constructed body, large wings, and very large hind feet, whence its specific name; besides these peculiarities it is also distinguished from every other Australian Bat by the hoary colouring of its fur, particularly on the lower part of the abdomen, where it is nearly white; it appears, however, subject to some variation in this respect, as in one of my specimens the hoary tint gives place to a pale reddish hue; but I believe hoary to be the prevailing colour.

General tint of the fur greyish-brown, becoming hoary on the posterior parts of the body, especially on the lower part of the abdomen, whence it gradually becomes paler, and fades into buffy-white on the vent; wing-membranes light brown.

The figures are of the size of life.

VESPERTILIO TASMANENSIS.

J. Gould and H. C. Richter del. et lith   Hullmandel & Walton Imp.

VESPERTILIO TASMANIENSIS.

Tasmanian Bat.

Noctulina Tasmanensis, Gray, List of Mamm. in Coll. Brit. Mus., p. 194.

It would appear that this species enjoys an unusually wide range of habitat; for not only does it inhabit Tasmania, but, according to Mr. Tomes, it is also found in the Philippines, and even on the continent of India. Had I not known that Mr. Tomes had closely investigated the Vespertilionidæ, and that from his intimate knowledge of the subject he is considered an authority in such matters, I should have hesitated to make this statement.

On submitting my drawings to Mr. Tomes, he suggested that the ears should be a little more indented on the lower side, after the manner of the Notch-eared Bat of Europe; but the Plate having been printed, this could not be attended to.

The specimen from which my figure was taken is in the British Museum.

The fur of this species is of a light brown hue, with a slight tinge of olive, and is lighter on the under than on the upper surface; the wing-membranes and the interior of the ear are of the usual purplish-brown hue; the nose and lips reddish flesh colour.

The figure is of the natural size.

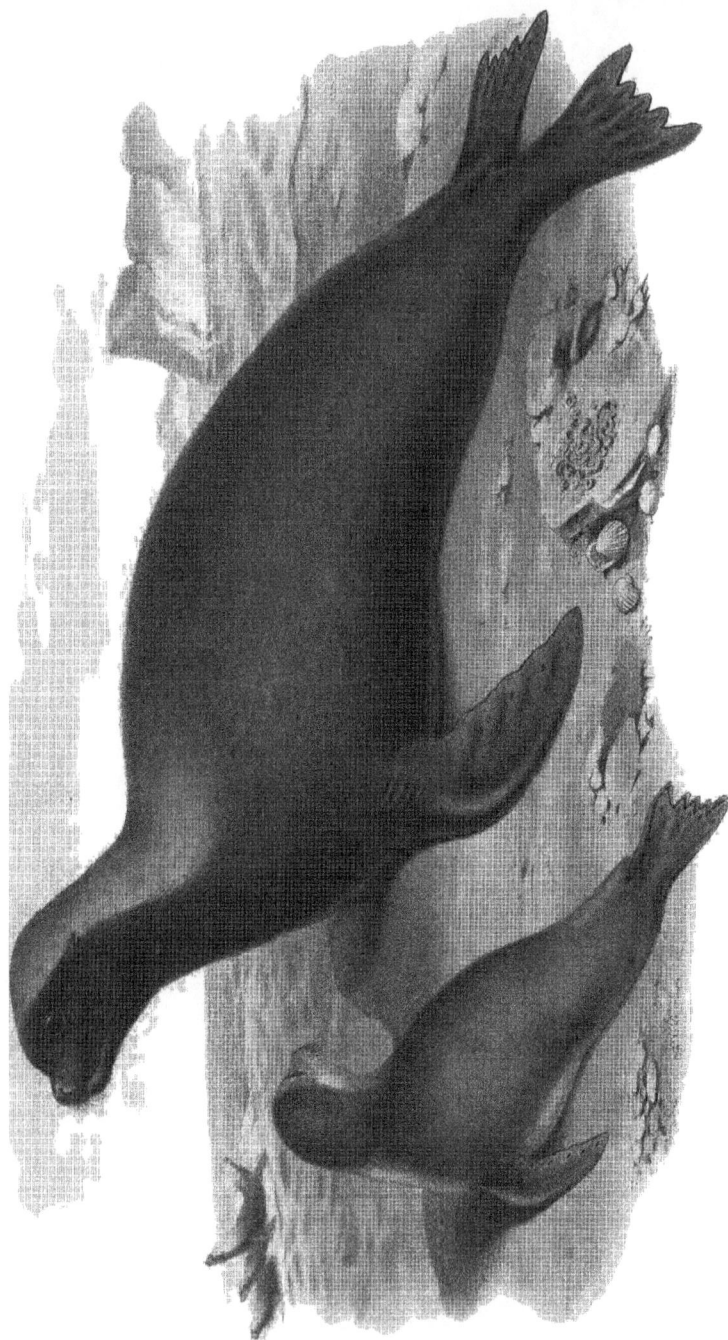

ARCTOCEPHALUS LOBATUS.

ARCTOCEPHALUS LOBATUS.

Cowled Seal.

Otaria cinerea, Gray in King's Narrat. Australia, vol. ii. p. 413.—Id. in Griff. Anim. Kingd., vol. v. p. 183 (not
Péron?), 1827.
Arctocephalus lobatus, Gray, Spic. Zool., i. t.   (skull).—Bull. Sci. Nat., vol. xvi. p. 113.—J. Brookes's Cat. Mus.,
p. 37, 1828.—Gray, Zool. of Ereb. and Terror, Mamm., pl. 16, p. 4.—Id. Cat. of Spec. of Mamm. in
Coll. Brit. Mus., part ii., Seals, p. 44.—Id. Proc. of Zool. Soc., part xxvii. p. 110.
Phoca lobata, Fisch. Syn., vol. ii. p. 574.
Otaria Lamairii, J. Müll. Wieg. Archiv, 1843, p. 334?
Otaria stelleri, (Mus. Leyden, 1845) Faun. Japon., t. 21, 22, 23 (animal), t. 22. fig. 3 (skull).
Otaria jubata, part, Gray, Cat. of Osteol. Coll. of Brit. Mus., p. 33.

THERE is perhaps no one group of the Mammals of Australia so little understood as the Seals; hence it is very gratifying when we are able to obtain any reliable information respecting the species that visit the rocky shores of that continent and the adjacent islands. As I did not see many of these animals during my visit to Australia, I must content myself with letting those who have say what they know of the subject, taking care that the animals are correctly figured, and that the passages quoted are correctly applied. I would also remark that the list of synonyms are given on the authority of Dr. Gray's ' List of the Seals contained in the Collection of the British Museum;' and as this gentleman has paid much attention to the Seals of the Southern Ocean, I have no doubt that they may be depended upon.

The specimens spoken of by Mr. Gilbert, in the note from his MSS. given below, as having been procured by him on the Houtmann's Abrolhos, as well as the one which Mr. Angus mentions as killed by Sir George Grey in Rivoli Bay, are all in the British Museum; and it is from these specimens that my figures are taken. There is but little doubt in Dr. Gray's mind that Mr. Gilbert's specimens from the Houtmann's Abrolhos are the female or young of the much larger male shot by Sir George Grey in Rivoli Bay, although the latter is twice the size of the former, being fully ten feet in length and as large in girth as a moderate-sized horse. No great length of time has elapsed since the islands in Bass's Straits and the south coast of Australia were first visited by the sealers; but in that comparatively short interval they have dealt out destruction among these inoffensive animals to such an extent that they are now all but exterminated. Collins (in 1798, when his account of New South Wales was published) mentions that " The rocks towards the sea were covered with Fur-Seals of great beauty, of a species which seemed to approach nearest to that known to naturalists as the Falkland Island's Seal." Few, if any, are now to be seen there.

" In the collection I now send you," says Mr. Gilbert, " you will receive eight Seals, of various sizes, the largest of which is a mature male, though it is not so large, by a third, as the very old ones, of which I saw several, but could not obtain either of them. Among them is a half-grown male and a full-grown female; the others are young animals, and the smallest a suckling.

" This animal is extremely numerous on all the low islands of the Houtmann's Abrolhos, particularly those having sandy beaches; but it does not confine itself to such places, being often found on the ridges of coral and madrepores, over which we found it very painful walking, but over which the Seals often outran us. On many of the islands they have been so seldom (perhaps, indeed, never before) disturbed, that I frequently came upon several females and their young in a group under the shade of the mangroves; and so little were they alarmed, that they allowed me to approach almost within the reach of my gun, when the young would play about the old ones, and bark and growl at us in the most amusing manner; and it was only when we struck at them with clubs that they showed any disposition to attack us, or defend their young. The males, however, would generally attack the men when attempting to escape: but, generally speaking, the animal may be considered harmless; for even after being disturbed they seldom attempt to do more than take to the water as quickly as possible. They differ much in colour, the males being considerably darker than the females."

I am indebted to Mr. G. F. Angus for a drawing of this animal, taken from the specimen killed by Sir George Grey, as mentioned above.

" I send you," says Mr. Angus, " a sketch of the Seal killed by Sir George Grey, while Governor of South Australia, in Rivoli Bay, on the south-east coast of that colony. I was with Sir George when it was shot and afterwards clubbed, and made my sketch, and took its admeasurements on the spot after death."

Dr. Gray states that this species and the *A. Hookeri* " are called Hair-Seals by the scalers because they are destitute of any under fur; but this appears to be the case only with the older specimens, for the young of *A. lobatus* is said to be covered with soft fur, which falls off when the next coat of hair is developed. The under fur is entirely absent in the half-grown *A. lobatus* in the British Museum collection."

The adult has the face, front and sides of the neck, all the under surface, sides, and back dark or blackish brown, passing into dark slaty grey on the extremities of the limbs; the hinder half of the crown, the nape and back of the neck rich deep fawn-colour; eyes black.

In the young a reverse of this colouring occurs, the upper surface being dark, and the face and under surface buff.

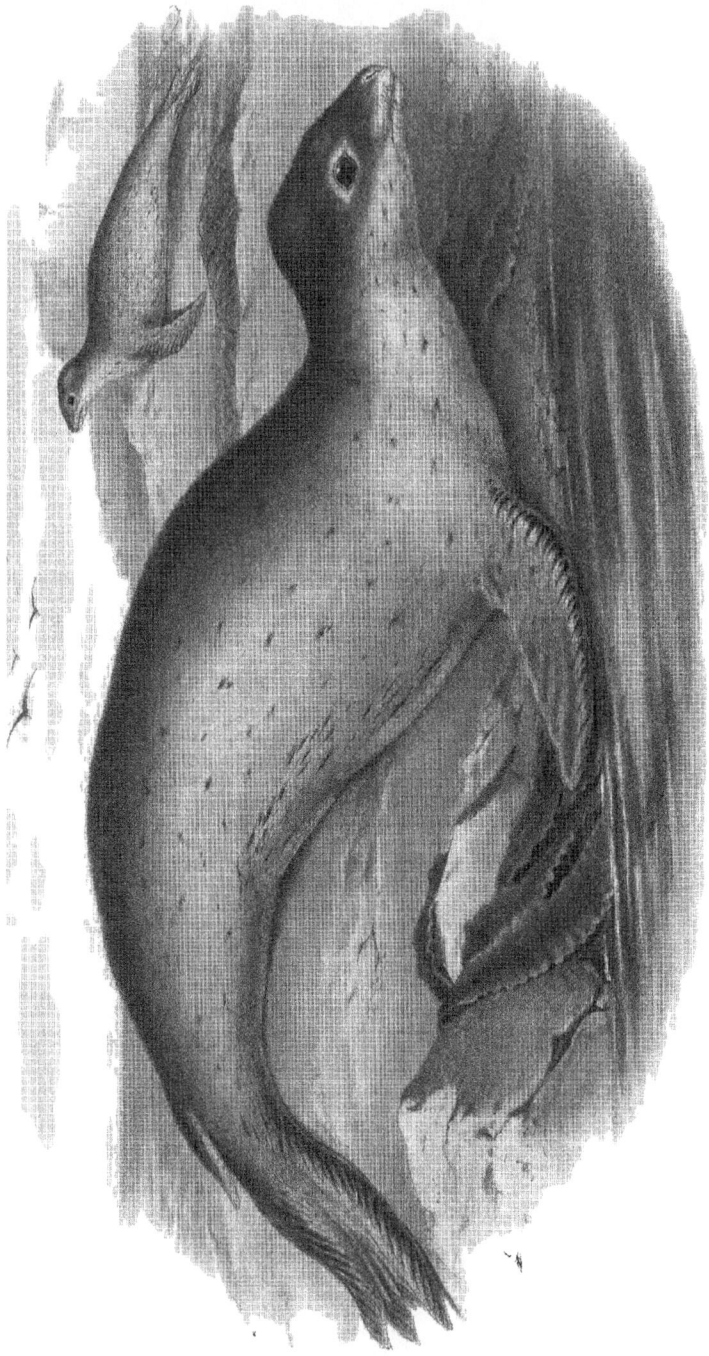

STENORHYNCHUS LEPTONYX.

STENORHYNCHUS LEPTONYX.

Sea Leopard.

Phoca Leptonyx, Blainv. Journ. Phys., vol. xci. p. 288, 1820.—Desm. Mamm., p. 247.—Cuv. Oss. Foss., vol. v. p. 208.
   t. 18. fig. 2.—Gray in Griff. Anim. Kingd., vol. v. p. 178.—Blainv. Ostéogr., Phoca, t. 1, and t. 4. fig. .
   (skull).—F. Cuv. Dent. Mamm., p. 118. t. 38 A.
Seal from New Georgia, Home, Phil. Trans. 1822, p. 240. t. 29 (skull).
Phoque quatrième, Blainv. in Desm. Mamm., p. 243 (note).—Cuv. Oss. Foss., vol. v. p. 207.
Stenorhynchus Leptonyx, F. Cuv. Dict. Sci. Nat., vol. xxxix. p. 549. t. 44.—Ib. Mém. Mus., vol. xi. p. 190. t. 13. fig. 1.
   —Ib. Dent. Mamm., p. 118. t. 38 A.—Nilsson, Wieg. Archiv, vol. vii. p. 307.—Ib. Scand. Faun., t.   .—
   Gray, Zool. of Ereb. and Terror, Mamm., t. 3 (animal), t. 4 (skull) p. 4.—Ib. Cat. of Osteol. Spec. in
   Brit. Mus., p. 31.—Blainv. Ostéogr., Phoca, t. 5. fig. 9 (teeth and skull).—Gray, Cat. of Spec. of Mamm.
   in Brit. Mus., part ii., Seals, p. 13.
Phoca Homei, Less. Dict. Class. Nat. Hist., vol. xiii. p. 417.
The small-nailed Seal, Hamilton Smith in Jard. Nat. Hist. Mamm., vol. viii. p. 180. t. 11.
Stenorhynchus aux petits ongles, Homb. et Jacq. Voy. à Pole Sud, t. 9.
Phoca ursina, or *Sea Bear*, Polack.
Sea Leopard of the Whalers.

Upon landing on the sandy beach of one of the quiet bays of Port Arthur, Tasmania, I found myself between the salt water and a huge specimen of the Seal figured on the accompanying Plate; of course, as I had never seen the animal before, it was not to be lost without a struggle; and, after a slight resistance on the part of the animal, a strong cord was fastened round its neck, with the view of towing it after my boat and killing it by drowning, that the specimen might not be injured; but the attempt at dispatching the animal by this means proved futile, as the more it was towed through the water, the more it appeared to gain strength, and other means of depriving it of life had to be resorted to. I have notices of two other specimens having been taken on the south coast of Australia, almost in the immediate neighbourhood of Sydney. For the particulars of their capture, as well as for a very fine drawing of the species, I am indebted to Mr. G. F. Angas, who made the latter immediately after the death of one of them. I mention these solitary instances of its occurrence, because I have reason to believe that the animal is not common in the localities mentioned.

The note accompanying Mr. Angas's drawing is somewhat interesting, inasmuch as it informs us that the stomach of the Seal contained a specimen of that remarkable animal the Ornithorhynchus.

"We have lately added to our Museum Collection," says Mr. Angas, "a fine specimen of an adult Sea Leopard (*Stenorhynchus Leptonyx*), killed some miles above the salt water in the Shoalhaven River; it had an Ornithorhynchus in its stomach when captured; it is much larger than one killed on Newcastle Beach. The dentition is exactly the same as that of the animal figured under the name above-mentioned in the 'Zoology of the Voyage of the Erebus and Terror.'"

I am again obliged to remark that the above list of synonyms is given on the authority of Dr. Gray. For my own part, I have not been able to give sufficient attention to the subject to vouch for their correctness, but Dr. Gray's well-won reputation will be a sufficient guarantee in this respect.

This species of Seal is of a more lengthened or slender form than the *Arctocephalus lobatus*; its length is about ten feet, and its weight probably four hundred pounds.

The general colouring of the animal is greenish creamy white, becoming of a dark slaty hue on the head and back, and speckled with the same dark hue on the sides.

CANIS DINGO, *Blumenb*

J. Gould and H.C. Richter, del. et lith.

Hullmandel & Walton, Imp.

CANIS DINGO, *Blumenb.*

The Dingo.

HEAD, OF THE SIZE OF LIFE.

THE opposite life-sized head of the Dingo, or native Australian Dog, is portrayed so faithfully, through the talent of Messrs. Richter and Krefft, that I am certain no one will regret my giving two plates of this animal: whether the head be viewed as a zoological illustration or as a work of art, it must be equally acceptable.

The natural history of the Dingo is so fully entered into in the letter-press accompanying the reduced figures, both from my own observation of the animal in a state of nature and from the writings of previous authors, that to recapitulate them here would be superfluous; I therefore refer my readers to that account.

It will be seen that the animal is subject to much variety of colour; I might therefore have multiplied the plates to almost any extent; but such a measure would have been of very questionable utility: I have therefore confined myself to one representing the normal style of colouring.

CANIS DINGO, *Blumenb.*

The Dingo.

Canis Dingo, Blumenb.—Shaw, vol. i. pl. 76.—Gray, List of Spec. of Mamm. in Coll. Brit. Mus., p. 57.
———— *familiaris*, var. *Australasiæ*, Desm.—Benn. Gard. and Menag. of Zool. Soc. del., vol. i. p. 51, with fig.
Chrysæus Australiæ, Lieut.-Col. Hamilton Smith in Jard. Nat. Lib. *Dogs*, vol. i. p. 188. pl. 10.

———————

"WHETHER the numberless breeds of dogs, which are the companions of the human race in every region of the globe, were originally descended from one common stock, and owe their infinite varieties solely to their complete domestication, the modifications by which they are distinguished having been gradually produced by the influence of circumstances,—whether, on the contrary, they are derived from the intermixture of different species, now so completely blended together as to render it impossible to trace out the line of their descent,—and whether on either supposition the primæval race or races still exist in a state of nature, are questions which have baffled the ingenuity of the most celebrated naturalists. Theory after theory has been advanced, and the problem is still as eagerly debated as ever, and with as little probability of arriving at a satisfactory conclusion. In the investigation of this difficult subject, however, as in the search after the philosopher's stone, many curious facts have been brought to light which would otherwise in all probability have remained buried in obscurity; and the causes which are continually operating to produce a gradual change of character, both in outward form and in intellectual capacity, among the brute creation, have received considerable elucidation. It is thus that theories, however erroneous in themselves, are frequently made subservient to the advancement of science, by the important facts which are incidentally developed by their authors in the ardour of their zeal for the establishment of a favourite hypothesis."

Such are the words of the late Edward Turner Bennett at the commencement of his paper on the history of the Dingo in "The Gardens and Menagerie of the Zoological Society delineated." Agreeing with Mr. Bennett in the impossibility of arriving at a satisfactory conclusion on the subject, I feel that I cannot close the present work without giving a figure and description of an animal which forms so prominent a feature in the fauna of Australia. It may be expected also that I should myself have formed some opinion as to its claim to be regarded as indigenous or otherwise; and if this opinion should be at variance with those of some Australian zoologists who have lately written on the subject, I may state that it has not been formed without due consideration. Without going into the probable origin of this particular race of dogs, or offering reasons why it should not be considered as indigenous, I may briefly state that I believe it has followed the black man in his wanderings from Northern Asia through the Indian Islands to Australia, the southern portion of which country appears to be its boundary in this direction; for I believe it has never been found in Van Diemen's Land in the wild or semi-wild state in which it occurs on the Australian continent. From what I saw of the animal in a state of nature, I could not but regard it in the light of a variety to which the course of ages had given a wildness of air and disposition; indeed it appeared to have all the habits of a skulking low-bred dog, and none of the determined air and ferocity of disposition of the wolf or jackal: in confirmation of this opinion, I may cite the facility with which the natives bring it under subjection, and the parti-colouring of its hairy coat; for although the normal colouring is red or reddish sand-colour; black, or black and white, individuals are not unfrequently seen; and that this variation in the colouring is not due to crossing with the domesticated races introduced when the country was first discovered, is proved by the following passage in the Appendix to "Collins's Voyage," a work published soon after the colonization of New South Wales, where he says, "the dogs of this country are of the jackal species; they never bark, are of two colours, the one red, with some white about it, the other black: some of them were very handsome." The existence of parti-coloured Dingos is still further confirmed by Mr. Gilbert's note on the animal, as observed by him in Western Australia: "The Dingo is very common over all parts of this colony. There are a very great number of varieties, varying from reddish brown to black, white, light brown, and black and white." Now, on the other hand, it may be affirmed that late geological discoveries will set aside the idea of its being a mere variety and tend to prove that this dog existed in Australia even prior to the aborigines; for it is said that a skeleton of a Dingo has been discovered at Warnamborl, beneath a bed of volcanic ash; but I believe no fossil remains have yet reached this country. The following letter on the subject has been kindly transmitted to me by Mr. Gerard Krefft, a gentleman to whom I am indebted for a beautiful drawing of the head, and an entire figure of the animal sketched either from life or immediately after it was killed:—

"In reply to your inquiry about the Australian Native Dog, I beg to state that it is proved without a doubt, as far as my own judgment goes, that the Dingo is an original inhabitant of the Australian continent.

"There is now, at the Museum in Melbourne, a fossil skull, found with other animal remains in a cave at

Mount Macedon, by Mr. Selwyn, the Geological Surveyor of Victoria. This skull, according to the authority of Professor M'Coy, is identical with that of the Dingo of the present day.

"An article to this effect was published by the learned Professor in the 'Argus' of 1857; but as it is not in my power to consult a file of this Journal, I am unable to furnish any further particulars.

"All the specimens of the Dingo procured by me during my stay at the Lower Murray were distinguished by a white tip at the extremity of the tail, and among the 'trophies' which so generally ornament shepherds' huts in Australia, I do not recollect to have seen a single tail without the white tip.

"The black variety is more scarce; the single specimen which I secured was a young bitch, quite black, except the inside of the fore legs and paws and the outside of the hind legs and paws, which were of a tan-colour. The head was more pointed than in the yellow variety, and had a distinct patch of white, about the size of a shilling, on each cheek.

"I made a drawing of the animal on the spot, and another one of the head, life size; both sketches are now, I trust, in the hands of Professor M'Coy. This dog had been prowling about Jamieson's Station for several nights; it fell at last a victim to strychnine, and I secured its skin."

During my wanderings in Australia I saw much of the Dingo in a state of nature, and can bear testimony to its great tenacity of life and the consequent difficulty of destroying it. I also witnessed the destructive nature of its habits in various ways, particularly its mode of "rushing" the sheep-fold, when it not only wantonly kills great numbers, but scatters the remainder to such an extent as almost to occasion the loss of the entire flock. It is not altogether for the purpose of supplying the cravings of hunger that the Dingo visits the sheep-pen, but in mere wantonness, dealing out its vengeance right and left with a single bite, which, although not fatal at the moment, the sheep seldom recovers, but lingers and soon dies. Mr. Gilbert states that its more usual mode of attack is to follow a flock of sheep, and when a lamb drops behind to immediately pounce upon and carry it off; and Collins mentions that such is its invincible pre-dilection for poultry, that not even the severest beatings can repress it.

"The Dingos, or native dogs, 'Warragal' of the Aborigines," says Dr. Bennett, " are the wolves of the colony, and are perhaps unequalled for cunning. These animals breed in the holes of rocks: a litter was found near Yas Plains, which the discoverer failed to destroy, thinking to return and catch the mother also, and thus exterminate the whole family; but the 'old lady' must have been watching him, for on his return-ing a short time after, he found all the little dingos had been carried away, and he was never able, although diligent search was made in the vicinity, to discover their place of removal. The cunning displayed by these animals, and the agony they can endure without evincing the usual effects of pain, would seem almost incredible, had it not been related by those on whose testimony every dependence can be placed. The following are a few among a number of extraordinary instances. One had been beaten so severely, that it was supposed all the bones were broken, and it was left for dead. Upon the person accidentally looking back, after having walked some distance, his surprise was much excited by seeing ' master dingo rise, shake himself, and march into the bush, evading all pursuit. One supposed to be dead was brought into a hut, for the purpose of undergoing 'decortication;' at the commencement of the skinning process upon the face, the only perceptible movement was a slight quivering of the lips, which was regarded at the time as merely muscular irritability: the man, after skinning a very small portion, left the hut to sharpen his knife, and returning, found the animal sitting up, with the flayed integument hanging over one side of the face. Another instance was that of a settler, who, returning from a sporting exhibition with six kangaroo dogs, met with a dingo which was attacked by the dogs and worried to such a degree, that finding matters becoming serious, and that the worst of the sport came to his share, the cunning dingo pretended to be dead; thinking he had departed the way of all dogs, they gave him a parting shake and left him. Unfor-tunately for the poor dingo, he was of an impatient disposition, and was consequently premature in his resurrection; for before the settler and his dogs had gone any distance, he was seen to rise and skulk away, but at a slow pace, on account of the rough treatment he had received; the dogs soon re-attacked him, when he was handled in a manner that must have effectually prevented any resuscitation taking place a second time. The Dingo, like all dogs in a state of nature, never barks, but simply whines, howls, and growls, the explosive noise being only found among the dogs which are domesticated."

I cannot conclude this paper without stating that the Dingo affords considerable exercise and amusement to the Nimrods of Australia, who hunt it precisely as the fox is hunted in England, and for which it forms no mean substitute.

The size of the Dingo is about that of the English Fox-hound, but it is much lower on the legs. The accompanying Plates represent the head of the natural size, and the whole animal reduced.